Analytical Scientists in Pharmaceutical Product Development

Task Management and Practical Knowledge

Kangping Xiao

Registered Office
John Wiley & Sons, Inc., 111 River Street, Hoboken, NJ 07030, USA

Editorial Office
111 River Street, Hoboken, NJ 07030, USA

For details of our global editorial offices, customer services, and more information about Wiley products visit us at www.wiley.com.

Wiley also publishes its books in a variety of electronic formats and by print-on-demand. Some content that appears in standard print versions of this book may not be available in other formats.

Library of Congress Cataloging-in-Publication Data

Names: Xiao, Kangping, author.
Title: Analytical scientists in pharmaceutical product development : task management and practical knowledge / Kangping Xiao.
Description: First edition. | Hoboken : Wiley, 2021. | Includes index.
Identifiers: LCCN 2020019775 (print) | LCCN 2021019776 (ebook) | ISBN 9781119547822 (hardback) | ISBN 9781119547853 (adobe pdf) | ISBN 9781119547891 (epub)
Subjects: LCSH: Drug development. | Chemistry, Analytic.
Classification: LCC RM301.25 .X53 2021 (print) | LCC RM301.25 (ebook) | DDC 615.1/9–dc23
LC record available at https://lccn.loc.gov/2021019775
LC ebook record available at https://lccn.loc.gov/2021019776

Cover Design: Wiley
Cover Images: Courtesy of Kangping Xiao

Set in 9.5/12.5pt STIXTwoText by SPi Global, Pondicherry, India

Printed in the United States of America

SKY10020874_090120

In loving memory of my father, Xianglin Xiao

Contents

Preface

Everyone is a Scientist and a Manager

A competent analytical scientist in a fast-paced pharmaceutical product devel opment project team should be an expert in analytical chemistry. This analytical scientist should also have a more than cursory understanding of organic chemistry, physical chemistry, and statistics. Moreover, to play a leadership role in developing a new medicinal product that is chemically stable, the analytical scientist should be knowledgeable in formulation technologies, manufacturing processes and practices, and packaging materials and configurations. Last but not least, since the analytical scientist is responsible for the testing results and data integrity, he/she should understand the regulatory strategies and requirements, quality assurance demands, and compliance obligations.

However, being technically competent does not automatically land the analytical scientist on a leading role in the industry. Good scientists are not necessarily good team players by training. For instance, a scientist who has gone through Ph.D. studies is most likely honored with a doctorate by the uniqueness of his/her research, by his/her independence or self-sufficiency, and not necessarily by how much he/she can collaborate with other fellow students. Instead of following other people's directions or suggestions, a scientist may take pride in being able to solve puzzles on his/her own. In the case of an analytical scientist, whose nature possibly is very much detail-oriented, he/she can become so attracted by details and gets lost in the weeds. The analytical scientist may forget that the entire project team is waiting for the data, while he/she is working hard trying to separate two impurity peaks on a chromatogram, or, is working hard to sort the data in a flawless format before sharing the results with the team.

To move career upward, scientists should recognize that in industry, every scientist has to be a manager.

There is a common misunderstanding of what a manager is. When you ask a laboratory bench scientist if he/she is a manager, the answer is often something like: "No, I am not a manager because no one reports to me." Quite oppositely, the answer should be: "Yes, I am a manager." A manager does not always mean a person who guides people. Regardless of whether having responsibilities for people, a person has a lot of relationships to manage at a workplace. As a member of a project team, a scientist needs to consciously manage working relations with the immediate teammates, with colleagues from other teams or departments, and probably with the clients from external companies. Those relationships are all different depending on the roles, responsibilities, and positions. Therefore, in order to manage project activities well, regardless of whether he/she has people management responsibility, the scientist should pay attention to manage those different working relationships. One important note here is, however, that the aspect of management in this book is not regarding people management or team building. It is about task management. The professional working relationships should be managed based on the understanding of each other's work, by effective and efficient execution of responsible tasks, and by competently delivering satisfactory results. An analytical scientist should manage the analytical development tasks based on (1) a strong and broad knowledge of science, (2) a fair recognition of the obligations other departments have, and (3) a clear understanding of the impact the analytical development work on the operation of other team members or functions. As a leader in a project team, the analytical scientist should be able to see the big picture while paying attention to details, and has to manage the work priorities appropriately, keep timelines in mind, and meet project deliverables. Scientists all have curiosities, and that makes them good scientists. However, in industry, respecting project/task timelines is equally important as to following curiosities. Being able to balance the desire of pursuing scientific curiosity versus the duty of moving projects forward with results that are not perfect but acceptable, is a valuable skill for an analytical scientist to possess. Continuous improvement is an essential part of the life cycle management of a product, an analytical method, and a work process. Furthermore, besides knowing the importance of managing the working relations and managing tasks in hand, to effectively work with the co-workers, the scientist should be able to conduct effective communications; be able to articulate his/her thoughts, ideas, or rationales; be good at storytelling; and is willing to help establish a trustworthy, safe, less stressful working environment. All the above-mentioned aspects are part of a managership that a scientist should possess. In that sense, everyone is a manager.

In summary, a true leader in a product development environment should have solid scientific expertise and necessary soft skills to manage projects and activities. Only when scientists apprehend that they are on the path of product development together with other colleagues, where their work is affected by the workflow from the upper stream and their work also impacts the downstream team members or functions, can they become good team players or true leaders.

There are many books specializing in analytical chemistry, organic chemistry, liquid and gas chromatography, pharmaceutical product degradation chemistry, pharmaceutical product formulation, process technology, statistics for analytical scientists, interpersonal and communication skills, time management, project management, etc. Those books provide in-depth knowledge and serve as good tool books. However, sometimes it is not an easy task for a scientist who does not have a strong educational background in those specific fields to comprehend all the contents of the books. Sometimes people are looking for a quick answer to the question they have or the issue they encounter during their daily work, and do not have sufficient time to do in-depth learning of something they did not learn in school. A book that provides practical knowledge and insights that are sufficient for one to become a well-rounded scientific leader, a book that covers broad but critical aspects in analytical development, is much needed. The potential readers are Analytical Chemists, Pharmaceutical Scientists, Regulatory Affair Specialists, Graduate School Students in Chemistry related fields, Project Managers, Laboratory Managers, Consultants, and Business and Marketing coworkers who are interested in having some technical conversations with scientists.

Analytical development is a broad concept. Different companies, different product categories (e.g., prescription medicines, over-the-counter (OTC) medicines, nutritional products, dietary supplements, herbal medicines or phytomedicines), and different medicinal product development stages can have or request vast different organizational settings. The roles and responsibilities of the analytical scientists can be quite different between teams, departments, and companies. Not every analytical scientist can have an opportunity to experience product development from conceptual molecules to product commercialization. I have been working in a consumer health product development environment and have the opportunity to witness the majority of the process for the development of an OTC medicinal product.

Writing this book is largely based on the desire of sharing my learning. It is not intended to turn it into a textbook-like reading experience. I hope the contents of the book are interesting to read and are practically useful to the readers. The learning that I am sharing in this book is certainly not gained solely through my own thinking. I am forever grateful that ever since my childhood, I have been able to have friends, classmates, teachers, mentors, and colleagues that are kind, talented, devoted to their study, research and work, and open to share their knowledge. I spent my undergraduate years at Wuhan University, which has the most beautiful campus and is one of the top universities in Chemistry in China. During my Ph.D. study in the Analytical Chemistry Laboratory of Professor Yoshio Umezawa at The University of Tokyo (Japan), I was able to learn from the world-class researchers: Dr. Philippe Bühlmann, Dr. Sugawara Masao, Dr. Tohda Koji, and Dr. Seiichi Nishizawa. There I began to understand the true meaning of hard-working from doing research together with the students and professors behind the Akamon (Red Gate), the symbol of this broadly respected as the top institution of undergraduate education and graduate research in Japan. The research topic, Molecular

Recognition, enabled me to see the chemistry at the molecular level and allowed me to conduct a fair amount of organic synthesis. One of the compounds that I synthesized about 25 years ago is still being sold by several well-known chemical merchants. Although I was granted fellowship of the Japan Society for the Promotion of Science at the end of my Ph.D. study, for some reason I chose to come to the United States and began my postdoctoral research in Professor Merlin Bruening's laboratory at Michigan State University. There I learned the surface modification, synthesis, and characterization techniques. Two years later, I transferred to the University of Michigan to pursue my second postdoctoral study under the guidance of Professor Mark E. Meyerhoff. In his laboratory, the team studied various nitric oxide-releasing hydrophobic polymers that have potential applications as thromboresistant coatings for a large number of blood-contacting biomedical devices, such as coronary artery stents. I was there exploring different surface modification methods to make the polymers adhere to the stainless steel surfaces of the stents. Since then, I stepped into the area of the separation science on the surface, i.e., the chromatography. I used High-Performance Liquid Chromatography (HPLC) to monitor the release of chemicals from the polymers and was fascinated by this technique. From time to time, I find myself thinking about how many people have been able to make a living by doing chromatography, thanks to the curiosity of the Russian-Italian botanist Mikhail Tsvet had about the filter paper extraction in the 1890s! To my surprise, I found the knowledge I have obtained from the molecular recognition, and the surface characterization was so useful that the chromatography science did not seem foreign to me at all. Quite oppositely, I found the events happening on the HPLC column surface were so familiar when I thought about the molecular interactions among the stationary phases, mobile phases, and various components. The time spent in Michigan awarded me with two life-long friends, Dr. Wei Zhang and Dr. Wei Qin.

Working in the industry after my postdoctoral research made me realize that being a super-individual is not enough. Thanks to my mentor and manager at Schering-Plough (now Merck), Dr. Abu M. Rustum, who is extremely knowledgeable and open to share his wisdom, I was able to tackle extremely challenging chromatographic separations. Maybe I still had the momentum of an academic researcher that I was able to publish several method development papers on top ranking chromatography journals. However, without the guidance and help from my collogues and friends, Dr. Min Li, Dr. Jenny Drake, Dr. Jinjian Zheng, Dr. Bin Chen, Hui Liu, Fangzhu Liu, Yuan Xiong, to name just a few, the achievement would not be nearly as much. After four years of having fun with chromatographic method developments, the desire of wanting to see the whole story of the drug product development, from molecules to the shelves of Walmart, made me move to Johnson & Johnson Consumer Care (McNeil). There I had all the opportunities I could dream about to learn from colleagues from various functions, such as analytical development, formulation development, regulatory department, quality assurance, project management, etc. My professional growth continued at Bayer Consumer Health, and I built an analytical development team from a few people in 2010 to more than twenty team members in 2020, from a testing work force to a technical team that leads the drug product development from the chemistry point of view. Along the journey, I have been so blessed to be surrounded by diligent team members, most distinctly Jianhua Li, Dr. Murty Varanasi, Sejal Patel, Bert Minerley, Nancy Tang, Dr. Song Liu, and Dr. Qiuxia Zhao.

Finally, the completion of this book would not be possible without the love and support of my family. A very special thank you to my wife, Qinhang, to my sweet, gifted, and artistic daughters, Cindy and Amy, for their continued encouragement.

The opinions, practices, and conclusions expressed herein are solely my own and do not express the views or opinions of my previous employers, Schering-Plough (now Merck), Johnson & Johnson, nor of my current employer, Bayer.

4 April 2020

Kangping Xiao
New Jersey, USA

1

Pharmaceutical Development at a Glance

In the United States, there are two main categories of medicinal products: prescription medicines and over-the-counter (OTC) medicines. The prescription medicines are prescribed to patients through appropriate health-care practitioners, while the OTC medicines are available to the general public through consumers' self-prescription. Many OTC products are once prescription medicines that are later switched to be over the counter with approvals from the US Food and Drug Administration (FDA). Consumers select OTC medicines based on their knowledge of the medical symptoms; by reading the labels on the product packaging; through their experiences with the products; according to their perceptions about the brand, manufacturer, packaging, product quality; and via their likes or dislikes of the sensational aspects of the product. Attractive dosage forms and appearance, pleasant flavors, soothing tastes, convenience at the time of use, costs, etc. are significant attributes (selling points) of an OTC medicinal product.

The processes for developing prescription medicines and OTC medicines are quite different. Prescription medicinal product development must go through the following steps: Step 1, Discovery and Development; Step 2, Preclinical Research; Step 3, Clinical Research; Step 4, Registration and FDA Review; and Step 5, FDA Post-Market Safety Monitoring. OTC medicinal product development has more varieties. A pharmaceutical company can commercialize an OTC medicine after going through either most or just a few above steps of the development [1].

1.1 Prescription Medicinal Product Development

The development of prescription medicine is a time-consuming process that requires a pharmaceutical company to spend lots of resources in both infrastructure and workforce. From infrastructure point of view, the company has

Analytical Scientists in Pharmaceutical Product Development: Task Management and Practical Knowledge, First Edition. Kangping Xiao.
© 2021 John Wiley & Sons, Inc. Published 2021 by John Wiley & Sons, Inc.

to make a lot of investment to (1) build research and development (R&D) facilities with sufficient space that holds various kinds of state-of-the-art instruments and laboratory equipment; (2) build modern manufacturing facilities and the associated waste treatment capacity; (3) build robust and effective document management systems to safeguard proprietary information and secure data integrity to comply with laws and regulations; and (4) build digital/electronic capability for data acquisition, processing, storage, and sharing. From the staffing point of view, the company must be able to attract and retain talents. Besides all the personnel needed to manage and run the business, a number of intelligent scientists are needed to carry out an enormous amount of research/discovery work. The medicinal product development starts from massive and fundamental researches that sometimes are tedious during drug discovery, that hundreds of molecules are studied. The development work then moves on to the safety and efficacy evaluations of the selected molecules. Once human clinical trials are involved, stringent regulations are applied to the studies from the sourcing of raw materials to the release of clinical trial materials to hospitals. High-quality documentation and noncompromised data integrity are a must for clinical studies. If the candidate molecule is lucky enough to make itself into a final medicinal product, information regarding Chemistry, Manufacturing, and Controls of all the ingredients in, the manufacturing process of, and the analytical methodologies employed to ensure the quality of the product are put together and submitted to health authorities for approval, which itself is lengthy and costly. Averagely a new prescription medicinal product approval in the United States takes more than a decade if counted from Step 2, preclinical research. The overall time is much longer if including Step 1, drug discovery.

1.1.1 Active Pharmaceutical Ingredient (API) Development

An Active Pharmaceutical Ingredient (API) is the molecule that possesses a desired therapeutic effect. During the drug discovery, a promising molecule is identified and studied thoroughly for its chemical nature, potential interactions with a target protein receptor, and its safety and toxicity. Studies regarding how potentially the molecule will be absorbed, distributed, metabolized, and excreted from human bodies are carried out [1]. Knowledge gained from the drug discovery helps the pharmaceutical company to design the clinical studies and begin the Investigational New Drug (IND) application, a process must be gone through before clinical trials can be carried out.

Once upon a time, compounds extracted from traditional remedies were isolated, characterized, and their medicinal efficacies were studied. From those compounds, the working chemicals were identified, purified, and synthesized to become the so-called active pharmaceutical ingredients. Nowadays, those low hanging fruits have been long-gone. The modern drug discovery starts with much more in-depth fundamental research at molecular levels. For illnesses such as cancers, Alzheimer's, and cardiovascular diseases, the knowledge accumulated from the entire scientific community, not just within one or two pharmaceutical companies, regarding the mechanisms behind those diseases is shared and applied in creating new medicinal molecules.

Upon identification of an API molecule, a robust synthetic route has to be established to produce the API consistently with a high purity (e.g. within 99–101% w/w). The analytical laboratories in pharmaceutical or chemical industries equip themselves with many modern technologies to ensure a synthetic route is effective, efficient, economical, environmentally friendly, and most importantly, to ensure the manufacturing of the API material is consistently at a high quality. Those analytical technologies include, but not limited to, high-performance liquid chromatography (HPLC), gas chromatography (GC), ion chromatography (IC), inductively coupled plasma (ICP), mass spectroscopy (MS), combinations of MS with other detection mechanisms such as LC-MS, GC-MS, ICP-MS, Fourier transform infrared spectroscopy (FTIR), near-infrared spectroscopy (NIR), Raman spectroscopy, X-ray powder diffraction, thermal analysis such as thermogravimetric analyzer (TGA), differential scanning calorimetry (DSC), and gravimetric technique such as dynamic vapor sorption (DVS), etc.

1.1.2 Preclinical Research

Before testing a candidate molecule in humans, the pharmaceutical company must demonstrate that the molecule does not cause serious harm. Instead of demonstrating its efficacy in curing the target disease, *Do No Harm* is the first-thing-first in medicinal product development. There are two types of preclinical research: *in vitro* (Latin for *in the glass* or *in the test tubes*) and *in vivo* (Latin for *within the living*) studies. Preclinical studies can provide detailed information on dosing and toxicity levels and serve as the decision-making basis for the next step, i.e. clinical studies in humans.

The FDA requires that good laboratory practices (GLP) are to be followed for preclinical laboratory studies. The GLP regulations are defined in 21 CFR Part 58.1: "Good Laboratory Practice for Nonclinical Laboratory Studies." These regulations set the minimum basic requirements for study conduct, personnel, facilities, equipment, written protocols, operating procedures, study reports, and a system of quality assurance oversight for each study to help assure the safety of FDA-regulated products [1].

1.1.3 Clinical Research – Phase 1, Safety and Dosage

After a successful preclinical study, the pharmaceutical company will conduct extensive clinical researches on the interactions between the API molecule and human bodies. Usually, there are four clinical phases. Each phase has a different purpose.

The Phase 1 clinical study is for safety evaluation of the API and for exploring its potential dosage. The study usually involves 20–100 healthy volunteers, and the length of study is about several months. In some cases, people with specific diseases or conditions, such as certain types of cancer, can also participate in Phase 1 trials. The study is more focused on the safety aspect. Evaluation is carried out on the tolerable amount of the drug molecule and what kind of acute side effects the molecule may cause. The Phase 1 trials can also provide some preliminary information about the effectiveness of the drug molecule. Approximately 70% of the studies potentially make their way to the next phase [1].

1.1.4 Clinical Research – Phase 2, Efficacy and Side Effects

The volunteers who participate in Phase 2 studies are patients with the disease or condition for which the drug molecule is targeting to cure. In Phase 2 trials, a placebo, i.e. a mixture of all the ingredients without the API, is used as a comparison to explore the efficacy of the medical product. Although usually a few hundred patients are involved, a Phase 2 clinical trial still is focusing on the evaluation of the drug molecule's side effects. The number of volunteers is yet considered not big enough to conduct a definitive evaluation of benefits versus risks of the medicinal product. The length of study can range from several months to about two years. Approximately 33% of the studies move to the next phase [1].

1.1.5 Clinical Research – Phase 3, Efficacy and Monitoring of Adverse Reactions

A Phase 3 clinical trial, also known as a pivotal study, involves a few hundreds to a few thousands of volunteers. Efficacy is the focus of a Phase 3 study. Drug safety, however, remains a primary focus throughout this phase. Not only the number of volunteers is much larger, the duration of the trial is also longer, from one to four years, which makes it possible for the pharmaceutical drug developer to study long-term side effects and other uncommon and unwanted reactions between human bodies and the API molecule. Approximately 25–30% of the studies move to the next phase [1]. If we pause and do a simple calculation of the final ratio of the studies that survive from Phase 1 through Phase 3, $70 \times 33 \times 30\%$, we find a stunning failing rate of more than 90% of the clinical studies for a new prescription drug product development!

If the API molecule is one of the lucky survivals that make through Phase 3 clinical research, the pharmaceutical company can apply for the right to manufacturing and selling of the newly developed medicinal product. In the United States, the drug developer files a New Drug Application (NDA) to the FDA.

Detailed information and guidance regarding the NDA applications can be found on the FDA website. Since 1938, every new drug has been the subject of an approved NDA before US commercialization. The NDA application is the path through which a pharmaceutical company seeks approval from the FDA of a new drug product for sale and marketing in the United States. The data gathered during the animal studies (preclinical) and human clinical trials (Phase 1 through 3) are in the NDA. The contents of the NDA must be able to enable FDA reviewer to assess whether the drug is safe and effective in its proposed use(s) and whether the benefits of the drug outweigh the risks; whether the drug's proposed labeling is appropriate; and whether the formulation, manufacturing process, and analytical methods are suitable to ensure the drug's identity, strength, quality, and purity. The documentation required in an NDA is supposed to tell the drug's whole story, including what happened during the clinical tests, what are the ingredients of the drug, the results of the animal studies, how the drug behaves in human bodies, and how it is manufactured, processed, and packaged.

1.1.6   Clinical Research – Phase 4, Post-Market Safety Monitoring

However, the whole story is not complete yet. After approval of the new medicine, Phase 4 trials are carried out as part of the post-market safety monitoring [1]. Although FDA's oversight on Phase 4 studies may not be mandatory or as stringent as it is during the other clinical trial phases, many pharmaceutical companies choose to conduct the studies and submit the results to the FDA. The demands for monitoring real-life outcomes, public anxiety over the safety of existing medicines, and changing regulatory environments all play a role in motivating the pharmaceutical companies to continue the studies after the NDA approvals. Indeed, the spirit of continuous monitoring is a critical aspect of the professionalism of a pharmaceutical manufacturer. For example although Methylparaben (methyl p-hydroxybenzoate), one of the several alkyl esters, such as methyl, ethyl, propyl, or butyl, of p-hydroxybenzoic acid, is widely used as a preservative in pharmaceutical products of lotions, ointments, creams, and local anesthetic solutions, and in cosmetic products of makeups, shaving products, hair care products, moisturizers, and deodorants, it has been reported to be associated with allergic reactions. While no definitive verdict from the FDA whether or not Methylparaben should be considered safe [1], many companies are proactively looking into the removal of the parabens in their products.

In addition, a Phase 4 trial can be used to support publications and to substantiate a label claim.

1.1.7 FDA Approval of a Prescription Medicine

Once it receives and accepts an NDA submission for review, the FDA will spend 6–10 months to conduct a thorough review. The review team consists of experts in corresponding fields. FDA medical officers and the statisticians will review clinical data, pharmacologists will review the data from animal studies, and analytical scientists will review method validation and stability data. FDA inspectors will travel to clinical study sites to inspect and look for evidence of fabrication, manipulation, or withholding of data. At the end of the review, the review team issues a recommendation, and a senior FDA official makes a decision [1].

In cases where the FDA is in favor of approving the NDA, the FDA will work with the applicant to develop and refine prescribing information, which is referred to as "labeling." Labeling accurately and objectively describes the basis for approval and how best to use the medicine. It is not unusual for the FDA to require the applicant to address questions based on existing data or even conduct additional studies. In some cases, the FDA may organize a meeting of one of its advisory committees to get independent expert advice and to permit the public to make comments. These advisory committees include a patient representative that provides input from the patient perspective.

1.2 Over-the-Counter (OTC) Medicinal Product Development

OTC medicinal products are Generally Recognized As Safe and Effective (GRASE). The general public can self-prescribe and purchase an OTC medicine from the market without obtaining guidance from a medical doctor. In the United States, "The Consumer Healthcare Products Association (CHPA)" speaks for the OTC industry. CHPA is a more than a 100-year old national trade association that represents the leading manufacturers and marketers of OTC medicines and dietary supplements. According to CHPA, there are approximately 800 OTC active ingredients available today that constitute more than 300 000 OTC medicines in the health-care marketplace. They are sold in more than 750 000 retail outlets including pharmacies, grocery stores, convenience stores, mass merchandise retailers, etc. The Drug Facts label instructs consumers on how to properly choose and use them. OTCs treat many common ailments including pain, fever, cough and cold, upset stomach, and allergies [2].

The OTC medicinal product development takes either one of the two approaches: (1) market the newly developed product in the United States by complying with an FDA OTC drug monograph; (2) file a new drug application (NDA) or an abbreviated new drug application (ANDA) with the FDA to seek approvals. For some minor changes in formulation, the NDA owners can file a supplement NDA (sNDA) to seek FDA approval. Overall, compared to prescription medicinal product development, the OTC medicinal product development is at a much faster pace.

1.2.1 FDA Monograph System

The US FDA establishes monographs to set regulatory standards for marketing OTC medicines [1]. These standards provide the marketing conditions for some OTC drug products regarding the active ingredients, labeling, and other general requirements. Instead of setting a standard for individual products, the FDA categorizes OTC medicines based on therapeutic classes. For each category, OTC drug monographs are developed and published in the *Federal Register*. The OTC drug review was established to evaluate the safety and effectiveness of OTC medicines marketed in the United States before 11 May 1972. It is a three-phase public rulemaking process (each phase requiring a Federal Register publication) resulting in the establishment of standards (drug monographs) for an OTC therapeutic drug class. Advisory review panels are responsible for the first phase. The panels are in charge of reviewing the active ingredients in OTC medicines to determine whether these ingredients are safe and effective for use in self-treatment. They are also in charge of reviewing claims and recommending appropriate labeling, including therapeutic indications, dosage instructions, and warnings about side effects and preventing misuse. The agency publishes the panel's conclusions in the Federal Register in the form of an advanced notice of proposed rulemaking (ANPR). After publication of the ANPR, the FDA allows a period for interested parties to submit comments or data in response to the proposal. According to the terms of the review, the panels classify ingredients in three categories as follows:

- Category I: generally recognized as safe and effective for the claimed therapeutic indication;
- Category II: not generally recognized as safe and effective or unacceptable indications;
- Category III: insufficient data available to permit final classification.

The second phase of the OTC drug review is the agency's review of active ingredients in each class of medicines, based on the panel's review of ingredients, public comment, and new data that may have become available. The agency, in turn, publishes its conclusions in the Federal Register in the form of a tentative final monograph (TFM). After publication of the TFM, a period is allotted for interested parties to submit comments or data in response to the agency's proposal.

The publication of final regulations in the form of drug monographs is the third and last phase of the review process. The monographs establish conditions under which certain OTC drug products are generally recognized as safe and effective. After publication, a final monograph may be amended, either on the Commissioner's own initiative or upon the petition of any interested person. OTC drug monographs are continually updated to add, change, or remove ingredients, labeling, or other pertinent information, as needed.

In summary, the OTC drug monographs are kind of recipe books that cover acceptable ingredients, doses, formulations, and labeling. Upon finalization and implementation of an FDA monograph, companies can make and market an OTC medicine without the need for FDA preapproval. These monographs define the safety, effectiveness, and labeling of the marketed OTC active ingredients.

The FDA OTC monograph system is not to be confused with the United States Pharmacopeia (USP) and National Formulary (NF) monographs. The FDA monograph system is about the safety of the OTC medicines; the USP-NF monographs are about setting public standards for analytical testing procedures and quality requirements for APIs, excipients, and medicinal products, regardless of whether they are in the FDA monograph system or not. The mission of the USP is to improve global health through public standards and related programs that help ensure the quality, safety, and benefit of medicines and foods [3]. A USP monograph for a finished product includes the name of the active ingredient, the strength/potency, dosage form, analytical procedures, and the specifications. The USP monographs set public standards of the identification, purity, quality for ingredients and products, for the pharmaceutical companies to follow. Some common measurements such as dissolution, disintegration, microbiological testing, etc. have general chapters to ensure that the industry follows the common standards. The general chapters also include methodologies regarding data treatment and statistics. In short, the FDA monograph system is to facilitate fast development and marketing of OTC medicines with APIs that belong to the GRASE category; and the USP monograph system is to provide analytical methodologies and standards to ensure consistent quality control among the pharmaceutical industry.

1.2.2 New Drug Application Process for an OTC Medicinal Product

When a medicinal product does not comply with the FDA OTC monographs, and the company wishes to market the new product as an OTC medicine, the company has to go through the NDA process. One situation is that a company can market an OTC medicine that is identical in active ingredient, dosage form, strength, route of administration, labeling, quality, performance characteristics, and intended use, to a previously approved application. In this case, the company can apply for an Abbreviated New Drug Application (ANDA) with evidence to support that the to-be-marketed product is identical to the Reference Listed Drug (RLD). ANDAs do not contain clinical studies but include information that can support the new product's bioequivalence to the RLD [1].

1.2.3 Clinical Trials in OTC Product Development

It is somewhat counterintuitive that an OTC medicinal product needs clinical trials. The molecules are well-studied, and their therapeutic effects are public knowledge. However, fierce competition among the OTC medicinal products marketed by various companies pushes the manufacturers to be more innovative in dosage forms, in delivery paths, and in product sensations. To achieve even a subtle competitive edge, the companies seek new claims, new indications, and new health benefits from existing products. Clinical evidence is, therefore, a vital part of OTC medicinal product development. Based on the website of clinicaltrials.gov [4], which is a database of privately and publicly funded clinical studies, there are a variety of clinical trials for OTC medicinal products. The clinical studies range from Phase 1 to Phase 4, and the sizes of trials can range from several tens to several thousands of volunteers.

1.2.4 Prescription to OTC Switch

Prescription to OTC switch refers to OTC marketing of a product that was once a prescription drug product, for the same dosage form, population, and route of administration.

An efficacy supplement should be submitted to an approved NDA for a prescription product if the sponsor plans to switch the medicinal product covered under the NDA to OTC marketing status in its entirety without a change in the previously approved dosage form or route of administration. An NDA 505(b)(1) should be submitted if the sponsor is proposing to convert some but not all of the approved prescription indications to OTC marketing status. An original NDA 505(b)(1) or 505(b)(2) needs to be submitted if the sponsor plans to market either a new product OTC whose active substance, indication, or dosage form has never previously been marketed as OTC [1].

References

1 https://www.fda.gov.
2 https://www.chpa.org.
3 https://www.usp.org.
4 https://clinicaltrials.gov/ct2/home.

2

Analytics in Fast-Paced Product Development

As briefly mentioned in Chapter 1, the process of prescription medicinal product development can be lengthy and complicated. It is possible that one analytical scientist throughout his/her career only works in a specific area, such as drug discovery; or works with some specialized instrument, such as ICP-MS, or NMR (Nuclear Magnetic Resonance). The situation is different in consumer health care over-the-counter (OTC) medicinal product development. The development timelines can be as short as a few months and in general, are not longer than a few years. A project team of OTC medicinal product development consists of colleagues from various functional areas. The team can include members from Analytical Development, Formulation Development, Packaging Development, Regulatory Affairs, Medical Affairs, Scale-up and Technology Transfers, Clinical Operations, Toxicology, Biostatistics, Quality Assurance (QA), Marketing and Sales, Business Development, Manufacturing Plants and associated Quality Control Laboratories, Patent Department, Project Management, Contract Manufacturing Organizations (CMO), and Contract Research Organizations (CRO). The analytical scientists on the team can potentially participate in the product development from ideation to product commercialization, and also participate in postmarket product maintenance. A lead analytical scientist can play a key role in many activities such as evaluation of novel ingredients; assessment of formulation prototype stability; quality control testing of the prototypes made for consumer evaluations or products for clinical trials and/or registration stability; dossier preparation for filings by working with regulatory affairs; helping QA unit to get ready for health authority inspections; assisting roll-outs of products to different regions or countries; and supporting marketed product maintenance as part of life cycle management of the product. Within a project team, not only can the analytical scientists apply the knowledge of analytical chemistry but also can they learn from colleagues from other functional areas and obtain helpful soft skill through working with teammates that have vastly different scientific and

Analytical Scientists in Pharmaceutical Product Development: Task Management and Practical Knowledge, First Edition. Kangping Xiao.
© 2021 John Wiley & Sons, Inc. Published 2021 by John Wiley & Sons, Inc.

cultural backgrounds. It provides an excellent opportunity for analytical scientists to gain a wide range of knowledge. On the other hand, however, it creates a big challenge for analytical scientists since they have to do well not only in science but also in project management, and even in people management.

Analytical development (AD) is a common name of a department that consists of analytical scientists who perform methods development and laboratory testing for supporting product development. However, based on its daily job function, the word *Analytics* seems to be more appropriate. According to the definition in dictionaries such as Merriam-Webster and Oxford, analytics is about a systematic, logical analysis of data or statistics. A more expanded explanation of analytics can be found on Wikipedia, "Analytics is the discovery, interpretation, and communication of meaningful patterns in data. It also entails applying data patterns towards effective decision-making. In other words, analytics can be understood as the connective tissue between data and effective decision making within an organization." This Wikipedia interpretation of analytics actually summarizes well what an AD department performs within an OTC medicinal product development department.

Ironically, often the function of a research and development (R&D) analytical development is confused with that of a quality control unit. The perception such as the analytical development scientists are just doing sample testing in the laboratories is a misconception. The QC laboratories at the manufacturing sites are the clients of the R&D analytical development and the end users of the methods. An analytical development team in a fast-paced pharmaceutical product development can and should carry many more responsibilities. To summarize, an analytical development team should:

1) Lead product development from a chemistry point of view by providing insights regarding excipient compatibility and product chemical stability based on scientific knowledge and experimental data;
2) Manage project activities, timelines, and deliverables, including collaboration with other functions, departments, and outside partners;
3) Develop state-of-the-art analytical methods for the analysis of assay, dissolution, blend/content uniformity, degradation and impurity, and for cleaning validation;
4) Carry out life cycle management of the analytical methods including validation, verification, transfer, and comparison; and help continuous optimization/improvement of the analytical procedures during routine operations at the QC laboratories;
5) Manage stability programs by either conducting in-house stability testing or outsourcing to contract laboratories (for which the responsibility of ensuring the quality of the testing data still resides in the analytical development);
6) Prepare analytical related Chemistry, Manufacturing, and Control (CMC) documentation for regulatory filing and respond to questions and requests from various health authorities regarding product quality and analytical methodologies; and

7) Ensure data integrity, documentation quality, and compliance of the work to the requirements of current Good Manufacturing Practices (cGMPs) (such as ensuring the good status of instrument qualification, calibration, and maintenance, laboratory house-keeping, and safety).

Therefore, an analytical scientist in a new product development team has duties much beyond just developing a high-pressure liquid chromatography (HPLC) method, testing samples, and providing data to the team. Analytical scientists should realize that instead of developing a perfect method, providing knowledge and direction to facilitate product development is the real focus of analytical development. Without being able to understand the potential interactions between the excipients and the APIs, without the ability to predict the chemical stability of the product, or without being able to provide interpretations of the obtained results unambiguously, an analytical scientist significantly hampers his/her potential of moving to leadership positions in the industry. Moreover, the analytical development team as a whole should possess a broad spectrum of talents. The team should have subject matter experts in Analytical Chemistry, Chromatography/ Separation Science, Organic Chemistry/Degradation Chemistry, Statistics, Spectroscopy techniques such as various infrared spectroscopy (mid-IR, near-IR), Raman, ultraviolet (UV), nuclear magnetic resonance (NMR), mass spectroscopy (MS), etc. Equally important, the team should have a good understanding of quality assurance, regulatory compliance, computerized software validation, data integrity, and if possible and actually is very much desired, a good understanding of formulation technologies and manufacturing processes.

2.1 Overall Development Process for New Products

Although different companies may have different workflows, and each product development has its uniqueness, the mission of pharmaceutical development is to design a quality product and its manufacturing process to consistently deliver the intended performance of the product [1]. The process illustrated in Figure 2.1 represents more or less a general course of an OTC medicinal product development.

The product development journey starts with an idea from an individual employee, or a result from a team brainstorming exercise, or a concept extracted from marketing insights, or a hidden pattern in consumer behaviors brought to light from the outcome of a survey, or a technology based on a new scientific finding or break-through, and so-on. The idea or proposal will be evaluated by various functions to ensure the potential product development endeavor has sound business viability, sufficient medical efficacy and safety, effective regulatory filing strategy, and technical feasibility. It will be beneficial if some preformulation

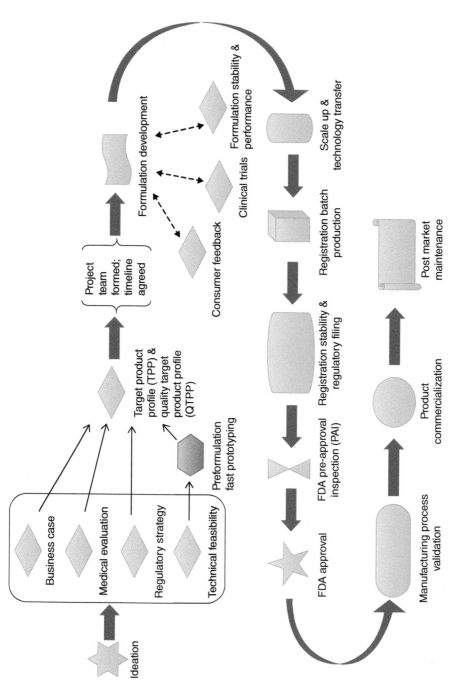

Figure 2.1 General workflow of over-the-counter medicinal product development.

work is conducted to strengthen the case, or to disapprove the proposal from the technical feasibility point of view. If the idea is considered worth pursuing, then the involved functions need to get together to determine the target product profile (TPP). As described in the Food and Drug Administration (FDA) draft guideline, the TPP embodies the notion of *beginning with the goal in mind* [2]. A team composed of higher level management or functional leaders specify the labeling concepts that are the goals of the drug development program; document the specific studies intended to support the labeling concepts, and then use the TPP to assist in a constructive dialogue with the FDA [2]. Speaking of labeling, it is not exaggerating to say that a large portion of a pharmaceutical product development endeavor is to ensure the correctness of the contents of and instructions on the product labels.

The TPP is not exactly the same concept as the quality target product profile (QTPP) that is described in International Conference on Harmonisation (ICH) Q8. The QTPP emphasizes more on the identification of Critical Quality Attributes (CQAs) of the product [1]. A CQA is a physical, chemical, biological, or microbiological property or characteristic of the drug substance, excipients, intermediates (in-process materials), and the drug product that should be within an appropriate limit, range, or distribution to ensure the desired product quality. Those characteristics can include aspects affecting product purity, strength, drug release, and stability. They can also include physical properties such as particle size distribution, bulk density, hardness, etc. Properly defining a TPP or a QTPP/CQA is imperative for the product and process development. The project team can modify the list of potential CQAs during the development with the increase of knowledge of the product and the manufacturing process. Quality risk management can be used to prioritize the list of potential CQAs for subsequent evaluation [3].

Once the product development goal is clear, a project (execution) team can formally form. A project manager is assigned, who kicks off the project, organizes the tasks, facilitates the communications among the team members, and monitors monetary spending and resource allocation, and measures project progress against agreed timelines. At the initiation of the project, the focus of the team is to come up with the best estimation of development costs, timelines, milestones, deliverables, and resource allocations. The development strategy such as regulatory filing strategy, marketing strategy, outsourcing versus in-house development, clinical trial designs, consumer feedback studies, stability programs, manufacturing site selections, establishment of a quality agreement with third party research or manufacturing organizations, etc. will be determined by and aligned among the team functional areas/members.

Formulation development officially starts after the project kickoff. An active pharmaceutical ingredient has to be formulated in certain dosage forms before patients or consumers can take it. Formulation development is a complex effort that involves the physiochemical characterization of the API, selection of compatible excipients, thorough characterization of the dosage form, and development of a reliable manufacturing process. Typical pharmaceutical dosage forms include oral tablets (swallowable, chewable, effervescent), powders, capsules, soft gels, chewable gels, liquids, syrups, suspensions, creams, lotions, ointments, inhalers, and injections for intravenous (IV), intramuscular (IM), or subcutaneous (SC) administration. The formulation scientists determine (1) the composition of a drug product, i.e. the formula, and (2) the steps and approaches of mixing the ingredients to form the product, i.e. the (manufacturing) process. Due to the nature of over-the-counter medicinal products, consumer desires and preferences are in the center of the product design. The product development is therefore not only science-based but also very much business-driven. Market research of consumer desires and performance of competitors provide crucial business insights. Evaluations of early prototypes by the potential or targeted consumers are essential parts of product development. Furthermore, clinical trials are sometimes necessary for developing a new product or coming up with a new claim, or even a new medical indication, of an existing product. This step is costly and time-consuming. Careful design of the consumer studies or clinical trials is very critical to achieving the outcomes that support the intended purposes.

As mentioned before, formulation development can be considered to include two aspects, a formula (*what*) and a process (*how*). The laboratory-scale formulation development is more focused on formula development to determine the composition of a product. The process development, i.e. the endeavors of establishing the most robust manufacturing process to consistently produce the proposed product, mainly occurs at the scale-up stage. It is not always necessary to separate formula development and scale-up development. The formulation scientists should have the knowledge and experiences to scale up the formulas they developed in the laboratory. Ideally, a manufacturing site has already been selected prior to the start of the project. The site expertise and equipment capability should have been confirmed much earlier before locking the formula. With the approaching of the finalization of scale-up processes, frequent communications between R&D and selected manufacturing sites become necessary. If an R&D has a pilot plant that can produce batches at 1/10 of the commercial scales, then the scale-up work can be done in house, which saves the time for product development, although not necessarily saves the maintenance cost for keeping a GMP facility within an R&D environment.

R&D involvement in the subsequent steps, such as registration batch production, preparation for FDA pre-approval inspections, manufacturing process validation, and product commercialization, is not as intensive. It is beyond the scope of this book to discuss the work involved in those product launch preparations and will thus be skipped.

Contributions of the analytical development start from the product ideation (Figure 2.2). The AD scientists should be actively involved in ideation and innovation, and should be able to contribute creative product ideas.

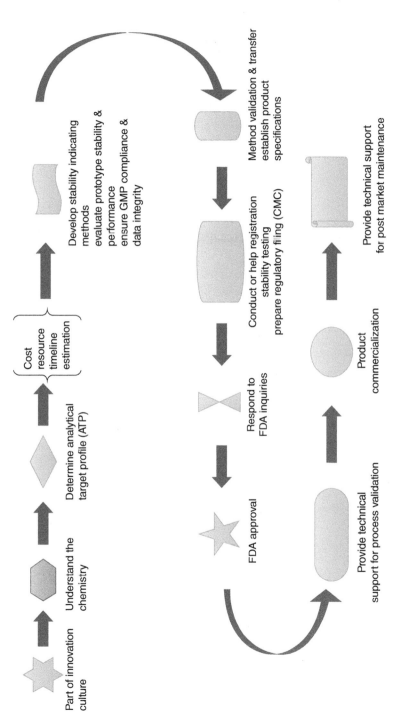

Figure 2.2 Contributions of analytical development in the overall workflow of over-the-counter medicinal product development.

The FDA holds the development of an OTC medicine to the same level of standards of the development of a prescription drug. Both medicines must be safe, effective, and at a high quality. To achieve those goals, scientists and specialists with various backgrounds work together, and each holds specific roles and responsibilities along the journey of product development. It is an analytical scientist's job to ensure high product quality by developing state-of-the-art analytical methods to analyze the potency and purity of the formula before consumers get their hands on the products. The challenge to an analytical scientist who works in the consumer health pharmaceutical industry, however, is that he/she must minimize the interferences from those pleasant-for-formula-but-extra-for-analytical ingredients added in the OTC medicinal products. Indeed, chromatographic method development for OTC medicines oftentimes is more challenging than method development for prescription drugs. As a side note, the analytical scientist's job becomes extremely challenging when OTC products contain natural ingredients, which can have dozens if not hundreds of compounds in their compositions. However, as mentioned at the beginning of this chapter, much beyond just providing testing results, analytical scientists make efforts to understand the chemistry of the proposed new formulas. They accumulate knowledge by conducting API forced degradation and excipient compatibility studies. The lead analytical scientists should be able to predict the stability and performance (such as dissolution and disintegration behaviors) of the proposed formulas based on previous knowledge and the results from stress studies. Although it is potentially a somewhat chicken-or-egg situation, i.e. having a method first or doing stress study first, the practical approach starts most likely with an approximately acceptable testing procedure followed by carrying out the forced degradation or compatibility studies. After having some understanding of the proposed formulas, the lead analytical scientists start to consider the analytical target profile (ATP) from the method development point of view. The concept of an ATP parallels the concept of a QTPP. The ATP defines the objectives of the analytical methodology applied at each stage of product development. In addition, the lead analytical scientists estimate the costs and resources needed to conduct the analytical work for supporting product development. Some negotiation skills possessed by the lead analytical scientists can help to get alignment on timelines, milestones, and deliverables among the team, rather than overload the analytical laboratory and potentially compromise the quality of the analytical work. The working relationship between the analytical development and the formulation development functions should be very close, transparent, and trust-based. Again, negotiation skills from the lead analytical scientists are needed to ensure the analytical laboratory work conducted is meaningful and insightful. During evaluation of formulation prototypes, the analytical scientists will use this opportunity to develop a fit-for-purpose analytical method. Beyond supporting the formulation prototyping, the analytical development scientists should work closely with the medical affairs and clinical operation colleagues to clearly understand the purpose of the consumer studies or clinical trials (if planned). For those studies, the analytical scientists should strictly follow the guidelines from health authorities (FDA, ICH, etc.) and follow company SOPs to ensure the analytical work conducted is scientifically sound, ethically right, and in compliance with cGMP regulations. At this stage, the analytical scientists should pay attention to the cGMP aspects of the work. Before testing the clinical trial materials, the analytical scientists should evaluate the method procedures against some typical validation parameters, such as specificity, linearity, accuracy, sensitivity, etc.

A labor-intensive aspect of the analytical work at the formulation process scale-up stage is that the analytical scientists may have to analyze a large number of samples for evaluations of blend uniformity and content uniformity. In the meantime, at this stage, the analytical procedures should be fine-tuned and polished to get ready for method validation and can be transferred to the manufacturing site(s). Some companies may have this practice that when transferring an analytical method to a manufacturing quality control laboratory, the transferring laboratory analytical scientist provides hand-on training to the receiving laboratory, which may be in a different country. This practice should be, in the author's opinion, the last resort for a method transfer. It is important to emphasize that being able to take a holistic approach to develop analytical methods and possessing a big-picture oriented mindset are essential for the lead analytical scientists. The best scenario is that only minimum method fine-tuning is necessary at this scale-up stage since at this moment the analytical scientists have to simultaneously cope with not only the heavy workload in the laboratory but also have to prepare a lot of documentation due to the upcoming regulatory filing (the CMC section). Moreover, the method developer should keep the end (final) users in mind from the beginning of the method development. The QC friendliness of the method should be one of the most critical aspects of the developed method. Therefore, it should not be a necessity for the transferring laboratory to physically send a scientist to train the receiving laboratory scientists on the nuances of the method. The polished analytical methods must be robust, easy to use, and straightforward in terms of experimental procedure and data processing. The method procedure should be written clearly and with sufficient details to enable the receiving laboratory scientists to carry out the test without formal hands-on training. Moreover, the manufacturing QC laboratories may desire that one method can be used for multiple similar products to enhance the laboratory work efficiency. The lead analytical scientists should have that request in mind from the initiation of the project and thus develop the methods accordingly. In summary, the lead analytical scientist should have the overall workflow in mind and think from the big-picture point of view, carry out work in a well-planned manner, and conduct frequent and effective communications with the colleagues from all other functions.

Depending on the companies, the R&D analytical development team may or may not get involved in the stability testing of the registration batches. If the analytical development laboratory has to generate data for registration stability, then the work must be conducted by following the cGMP regulation and the data integrity must be ensured. On the other hand, whether or not performing chemical testing for registration stability, the analytical development scientists will need to prepare or help prepare a lot of documents for regulatory filing. The method procedures, method validation/transfer reports, product specifications, justification of the specifications, reference standard certifications, stability analysis, dissolution method development report, etc. will need to be drafted and approved by the analytical development, regulatory affairs, and QA units. The importance of developing and implementing good documentation practices in analytical development is needless to say. Years after the regulatory approval, the FDA can question the method and the data, and the method developer and the scientists who generated the data may be asked to answer questions in front of the FDA auditors, during the FDA pre-approval inspection or general GMP inspections.

Once received the approval of the new product from the FDA, the manufacturing site needs to conduct manufacturing process validation. The involvement of the R&D analytical development will not be as much as that of before the product approval and may serve as a helper/consultant during the process validation. The product life cycle management, including the maintenance of the analytical methods, remains effective while the product is in the market until the product discontinues. The manufacturing site quality control laboratory will perform trend analysis to monitor the method performance. The R&D analytical development scientists may need to get involved in the method troubleshooting due to the observation of new impurities or degradation products, or due to the change of HPLC column surface chemistry. It is not rare that the HPLC column vendor happens to discontinue manufacturing the column that is in the method.

In the following sections, more detailed discussions regarding the roles and responsibilities of some other functions, and how the analytical development can effectively and efficiently work with those functions, are presented.

2.2 Regulatory Strategy and Analytical Development

The pharmaceutical industry is highly regulated, and it must obey regulations/laws and rules, must follow guidelines and general agreements. Since each country has its own health agency, regulation requirements seem to be proliferating rather than harmonizing across the globe. The followings are a few well-known agencies or international organizations:

- FDA
- EMA (European Medicine Agency)
- PMDA (The Pharmaceuticals and Medical Devices Agency, Japan)
- MHRA (The Medicines and Healthcare products Regulatory Agency, United Kingdom)
- NMPA (National Medical Products Administration, China)
- BfArM (The Federal Institute for Drugs and Medical Devices, Bundesinstitut für Arzneimittel und Medizinprodukte, Germany)
- ANVISA (The National Health Surveillance Agency, Agência Nacional de Vigilância Sanitária, Brazil)
- TGA (The Therapeutic Goods Administration, Australia)
- MSNA (The National Agency for the Safety of Medicines and Health Products, France)
- Health Canada
- ICH (International Conference on Harmonisation)
- WHO (World Health Organization)

Every pharmaceutical company must have a special function, the Regulatory Affairs, whose duty is to ensure that the company complies with the specific regulations of the countries where the company intends to market its products. Regulatory affairs professionals are the faces of the company when working with government regulatory agencies by presenting medicinal product registration documents. Within the company, regulatory affairs professionals determine the filing strategies for a medicinal product from as early as product development planning and actively engage in every step of the product development. Many times, the filing strategy defines the strategy of formulation development and analytical method development. Apart from preparation for regulatory submissions, regulatory affairs professionals are also responsible for post-approval maintenance to keep the products on the markets.

Due to the broad requirement in legislation regarding drug development, manufacturing, and distribution, regulatory affairs professionals come from various scientific backgrounds. No single major in a college can cover all the aspects a regulatory affairs department manages. Interestingly, many regulatory affairs professionals have previous experiences in analytical sciences.

The relationship between regulatory affairs and analytical development is very close. Many times when a health authority issues a deficiency letter to a company, the questions center around the analytical methodology and data integrities. Understanding the critical regulatory requirements is essential to the success of the analytical scientists who play leading roles in product development. The following sections will provide some highlights of several important FDA guidelines [4].

2.2.1 NDA and ANDA Filing

In the United States, the documentation required in a New Drug Application (NDA) is supposed to tell the whole story of the product and its development. The information includes what happened during the clinical tests, what are the ingredients, what are the results from the animal studies, how the drug behaves in human bodies, and how to manufacture, process, and package the drug. The FDA website has detailed guidelines [4]. For drug applications in Europe or other countries or regions, detailed application processes and requirements are on corresponding official websites of the health authorities. In the United States, the Abbreviated New Drug Application (ANDA) provides a shortened approval pathway for duplicates of a previously approved medicinal product, which is referred to as a Reference Listed Drug (RLD). A company may receive approval of an ANDA without submitting the same type and extent of information that is requested for a stand-alone new drug application to establish the safety and effectiveness of the proposed product. An ANDA must contain information to demonstrate that the proposed drug product and the applicable RLD are the same in terms of the active ingredient(s), dosage form, route of administration, strength, previously approved conditions of use, and, with certain exceptions, labeling. An ANDA must also include sufficient information to demonstrate that the proposed product is bioequivalent to the RLD, and to ensure the product's identity, strength, quality, and purity [4].

For filing, the application documentation should follow the Common Technical Document (CTD) format, which was developed by the International Council for Harmonisation of Technical Requirements for Pharmaceuticals for Human Use (ICH) in an attempt to streamline the submission requirements, especially for Japan, the European Union, and the United States. The CTD collects quality, safety, and efficacy information into a common format that has been adopted by ICH regulatory authorities. The electronic CTD is the standard format for electronic regulatory submissions, amendments, supplements, and reports to FDA's Center for Drug Evaluation and Research (CDER) and Center for Biologics Evaluation and Research (CBER) [4]. As of 5 May 2017, New Drug Applications (NDAs), Abbreviated NDAs (ANDAs), and Biologics License Applications (BLAs) must be submitted using eCTD format [4]. As of 5 May 2018, Commercial Investigational New Drug Applications (INDs) and Master Files must be submitted using eCTD format [4]. A CTD contains the following modules:

- Module 1: Administrative Information and Prescribing Information. It contains all administrative documents (e.g. application forms, claims of categorical exclusion and certifications), and labeling.

- Module 2: Summaries. It includes the summary documents such as overall CTD table of contents and introduction to the summary documents. The overviews and summaries contain the following additional documents:
  - Quality overall summary
  - Nonclinical overview
  - Clinical overview
  - Nonclinical summary
  - Clinical summary
- Module 3: Quality; see section below for details.
- Module 4: Nonclinical. It contains the nonclinical study reports and related information.
- Module 5: Clinical. It contains clinical study reports and related information.

For ANDAs, the applications generally do not contain data that are typically in Module 4. If nonclinical study reports or safety assessments are submitted in support of a proposed specification, i.e. toxicology studies to qualify (1) impurities per the ICH guidance for industry Q3A and Q3B, (2) residual solvents, (3) leachables, or (4) excipients, these reports or assessments are included in Module 4. Module 5 contains all of the clinical study report data needed to support the application and to demonstrate that the generic drug product is bioequivalent to the RLD.

2.2.2 Module 3 (CMC) of Common Technical Document

Among those modules, Module 3 is the central information body that needs significant contributions from analytical development. Module 3 contains the CMC information necessary to support the application. The once "Chemistry, Manufacturing, and Controls (CMC)" section has been renamed to "Pharmaceutical Quality/CMC" by FDA [4]. Any analytical procedure submitted should be described in sufficient detail to allow one skilled analytical scientist to reproduce the conditions and obtain results comparable to what is in the application. The following sections in Module 3 are directly related to the work of analytical development:

3.2.P.4 Contains information on the control of the *excipients*, including the identity of the source of inactive ingredients (including the supplier and the supplier's address) and the grades (e.g. compendial or noncompendial).

3.2.P.4.1 Contains the testing specifications, including the retest schedule and the excipient manufacturer's or supplier's COA(s).

3.2.P.4.2 Contains the analytical procedures for the noncompendial methods used for testing the excipients. For compendial excipients, applicants should reference the United States Pharmacopeia (USP) or the National Formulary but need not mention the analytical procedure.

3.2.P.4.3 Contains the validation data of the noncompendial or in-house analytical procedures.

3.2.P.4.4 Contains a justification of the specifications and includes (1) the applicant's or the drug product manufacturer's COA(s), (2) the residual solvents statement(s) from the manufacturer(s), and (3) the bovine spongiform encephalopathy, transmissible spongiform encephalopathy, and melamine certifications, as applicable.

3.2.P.5 Contains information on the control of the *drug product*.

3.2.P.5.1 Contains the specifications for the drug product, including the microbiological specifications (e.g. the microbial limits, sterility, and bacterial endotoxins), as applicable. These specifications should include the tests, acceptance criteria, and references to methods in a tabular format.

3.2.P.5.2 Contains a description of the analytical procedures (compendial and/or in-house) used for testing the drug product, including any microbiological tests, as applicable. For sterile drug products, this section should contain the methods for the product release tests (e.g. sterility tests or bacterial endotoxins tests [if applicable]).

3.2.P.5.3 Contains the validation of the analytical procedure, including

1) The full validation reports for the in-house methods and their equivalence to USP procedures, if available for the drug product.

2) A verification of USP General Chapter <1226> "Verification of Compendial Procedures," if referenced.

3) The legible spectra and chromatograms for reference standards and test samples.

4) The Sample Statement(s) of Availability and identification of the finished dosage form of the drug products. For sterile drug products, this section should contain the validation procedures and results for the microbiological analytical tests (e.g. sterility tests or bacterial endotoxins tests [if applicable]).

3.2.P.5.4 Contains the batch analysis, including the executed COAs for all presentations and/or strengths of the finished dosage form. The applicant should clearly identify the drug product batch(es) used in any BE studies, including the study identification number.

3.2.P.5.5 Contains the characterization of impurities. FDA recommends that applicants control all degradation products and process solvents if they are used during the manufacture of the finished dosage form. FDA also recommends that applicants complete the Summary Tables for the Listing and Characterization of Impurities and Justification of Limits in Drug Substance and Drug Products.

3.2.P.5.6   Contains the justification of the specifications, including but not limited to references to compendia (e.g. the USP or the Japanese Pharmacopeia), the ICH, and/or the RLD analysis. FDA recommends that applicants complete the Summary Tables for the Listing and Characterization of Impurities and Justification of Limits in Drug Substance and Drug Products.

3.2.P.6   Contains information about the reference standards or reference materials used for testing the drug product.

3.2.P.7   Container Closure Systems (not directly related to analytical development).

3.2.P.8   Contains the stability data

3.2.P.8.1   Contains the stability summary and conclusions for the finished dosage form, including

1) The pre-approval stability protocol.
2) The proposed expiration dating period for marketing packaging.
3) The proposed expiration dating period for bulk packaging, if applicable.
4) A storage temperature statement.

3.2.P.8.2   Contains the post-approval stability protocol. If the applicant and drug product manufacturer are different entities, stability protocols should be provided by the applicant. This section should also contain analytical procedures and testing schedules for maintenance of the microbial product quality (e.g. the container closure integrity/sterility, bacterial endotoxins, and microbial limits), as appropriate.

3.2.P.8.3   Contains stability data, including

1) Accelerated and long-term data.
2) Intermediate stability data, if applicable.
3) The batch numbers on the stability records that are the same as the test batch.
4) The date the stability studies were initiated.
5) The date that each stability sample was removed from the stability chamber for each testing time point.
6) Data on all presentations of the container-closure system. For primary batches of liquids, solutions, semi-solids, and suspensions, the product should be placed into worst-case

and non-worst-case scenarios. For post-approval stability studies, the applicant should pick the worst-case orientation for the study.

The following information and data can also be included in this section:
1) One-time special stability studies conducted to confirm the quality of the constituted drug products (for example, parenterals and/or powders reconstituted with diluents and/or drug admixtures) per the labeling's instructions
2) One-time thermal cycling studies (freeze-thaw/heat-cool), as applicable
3) One-time in-use stability studies for oral liquids and other dosage forms (e.g. a solution to be used within a certain period of opening the container per labeling instructions, compatibility with a dropper when provided as part of the container closure system), as applicable.

2.2.3   Supplements and Other Changes to an Approved NDA or ANDA

Companies are allowed to make changes to medicinal products or their labels after approval of the application. To change a label, market a new dosage form or a different potency/strength of a medicine, or change the way of manufacturing a medicinal product, a company must submit a Supplemental New Drug Application (sNDA) [5]. A supplement number is assigned and is associated with an existing FDA NDA number. The FDA must approve all significant changes in areas such as analytical testing procedures, packaging configurations, or product ingredients, to ensure the conditions set for the original product remain uncompromised.

If the holders of NDAs and ANDAs who intend to make post-approval changes in (1) components and composition, (2) manufacturing sites, (3) manufacturing process, (4) specifications, (5) container closure system, and (6) labeling, as well as (7) miscellaneous changes, and (8) multiple related changes, they must follow the FDA guidance for sNDA [5]. There are three reporting categories, a major change, a moderate change, and a minor change, that has a substantial, moderate, or minor potential, respectively, to have an adverse effect on the identity, strength, quality, purity, or potency of a drug product. Changes in the qualitative or quantitative formulation, including inactive ingredients, as provided in the already approved application, are considered major changes. A major change requires the submission of a Prior Approval Supplement. A moderate change can have two types of filing strategies. One type of moderate change requires the submission of a supplement to the FDA at least 30 days before the distribution of the drug product made using the change. This type of supplement is called a "Supplement - Changes Being Effected in 30 Days" (CBE-30). The drug product made with a moderate change cannot be distributed if the FDA informs the applicant within 30 days of receipt of the supplement that a prior approval supplement is required. If the FDA informs the applicant within 30 days of receipt of the supplement that information is missing, distribution must be delayed until the supplement has been amended to provide the missing information. The other type of moderate change is that the FDA may identify certain moderate changes for which distribution can occur when the FDA receives the supplement. This type of supplement is called a "Supplement - Changes Being Effected" (CBE). If, after review, FDA disapproves a changes-being-effected-in-30-days supplement or changes-being-effected supplement, FDA may order the manufacturer to cease distribution of the medicinal products made with the disapproved change. For minor changes, a description of the changes in the pharmaceutical company's next Annual Report can suffice. For example, deletion or reduction of an ingredient intended to affect only the color of the medicine may be reported in an annual report.

Areas need to be assessed due to the changes include (1) conformance to specifications; (2) impurity or degradation product profile, with toxicology tests to qualify a new impurity or degradation product or to qualify an impurity that is above a previously qualified level, when necessary; (3) physical properties such as the hardness or friability of a tablet after certain changes; (4) bioequivalence when required, which could include multipoint and/or multimedia dissolution profiling and/or an in vivo bioequivalence study; (5) extractables and/or leachables from new packaging components; and (6) moisture permeability of a new container closure system. Equivalence comparisons are required between the products made before and after the changes. Equivalent does not mean identical, and the equivalence comparisons frequently have a criterion for comparison with a calculation of confidence intervals relative to a predetermined equivalence interval.

All changes made in the specifications of an already approved application must be submitted in a prior approval supplement. *Specifications* (i.e. tests, analytical procedures, and acceptance criteria) are the quality standards provided in an approved application to confirm the quality of drug substances, drug products, intermediates, raw materials, reagents, components, in-process materials, container closure systems, and other materials used in the production of a drug substance or drug product. For example, a test, an analytical procedure, and an acceptance criterion can be an assay, a specific and fully described HPLC procedure, and a range of acceptable potency of 98.0–102.0%, respectively.

If an assessment indicates that a change has adversely affected the identity, strength, quality, purity, or potency of the medicine, a prior approval supplement is needed regardless of the recommended reporting category for the change. For example, a process change recommended for a CBE-30 supplement could cause the formation of a new degradation product that requires identification and/or qualification. The new degradation product has to be qualified to indicate that there are no safety concerns relating to the new finding. Indeed, the analytical development should have a proactive attitude and carry out the impurity/degradation product identification much earlier than finding an Out-of-Specification result. The lead analytical scientists should conduct the structure elucidation, coordinate customized synthesis of the newly identified compound if necessary, and work closely with the toxicology colleagues to qualify the new impurity/degradation product.

The analytical procedure in the approved application that is designated for use in evaluating a defined characteristic of the drug substance or drug product is called a regulatory analytical procedure. The analytical procedures in the US Pharmacopeia/ National Formulary (USP/NF) are recognized as the regulatory analytical procedures for compendial items. However, tests and associated acceptance criteria and regulatory analytical procedures in addition to those specified in the USP/NF may be required for approving compendial items. And, although the sNDA may include alternative analytical procedures to the approved regulatory analytical procedures for testing the drug substance and drug product, for purposes of determining compliance with the FDA "Federal Food, Drug, and Cosmetic Act," regulatory analytical procedures are used.

2.2.3.1 Major Changes – Prior Approval Supplement

The following are examples of changes in specifications considered to have a substantial potential to have an adverse effect on the identity, strength, quality, purity, or potency of a drug product as these factors may relate to the safety or effectiveness of the drug product.

1) Relaxing an acceptance criterion.
2) Deleting any part of a specification.
3) Establishing a new regulatory analytical procedure including designation of an alternative analytical procedure as a regulatory procedure.
4) A change in a regulatory analytical procedure that does not provide the same or increased assurance of the identity, strength, quality, purity, or potency of the material being tested as the regulatory analytical procedure described in the approved application.
5) A change in an analytical procedure used for testing components, packaging components, the final intermediate, in-process materials after the final intermediate, or starting materials introduced after the final intermediate that does not provide the same or increased assurance of the identity, strength, quality, purity, or potency of the material being tested as the analytical procedure described in the approved application except as otherwise noted. For example, a change from an HPLC procedure that distinguishes impurities to (1) an HPLC procedure that does not, (2) another type of analytical procedure (e.g. titrimetric) that does not, or (3) an HPLC procedure that distinguishes impurities but the limit of detection and/or limit of quantitation is higher.
6) Relating to testing of raw materials for viruses or adventitious agents:
   - relaxing an acceptance criterion;
   - deleting a test;

- a change in the analytical procedure that does not provide the same or increased assurance of the identity, strength, quality, purity, or potency of the material being tested as the analytical procedure described in the approved application.

2.2.3.2 Moderate Changes – CBE-30
The following are examples of changes in specifications considered to have a moderate potential to have an adverse effect on the identity, strength, quality, purity, or potency of a drug product as these factors may relate to the safety or effectiveness of the drug product.

1) Any change in a regulatory analytical procedure other than those identified as major changes or editorial changes.
2) Relaxing an acceptance criterion or deleting a test for raw materials used in drug substance manufacturing, in-process materials prior to the final intermediate, starting materials introduced prior to the final drug substance intermediate, or drug substance intermediates (excluding final intermediate) except as provided for in the point No. 6 "Relating to testing of raw materials for viruses or adventitious agents" in the Major Changes.
3) A change in an analytical procedure used for testing raw materials used in drug substance manufacturing, in-process materials prior to the intermediate, starting materials introduced prior to the final drug substance intermediate, or drug substance intermediates (excluding final intermediate) that does not provide the same or increased assurance of the identity, strength, quality, purity, or potency of the material being tested as the analytical procedure described in the approved application except as provided for in the point No. 6 "Relating to testing of raw materials for viruses or adventitious agents" in the Major Changes.
4) Relaxing an in-process acceptance criterion associated with microbiological monitoring of the production environment, materials, and components that are included in NDA and ANDA submissions. For example, increasing the microbiological alert or action limits for critical processing environments in an aseptic fill facility or increasing the acceptance limit for bioburden in bulk solution intended for filtration and aseptic filling.
5) Relaxing an acceptance criterion or deleting a test to comply with an official compendium that is consistent with FDA statutory and regulatory requirements.

2.2.3.3 Moderate Changes – CBE
The followings are examples of changes in specifications considered to fall within the category of CBE.

1) An addition to a specification that provides increased assurance that the drug substance or drug product will have the characteristics of identity, strength, quality, purity, or potency that it purports or is represented to possess. For example, adding a new test and associated analytical procedure and acceptance criterion.

2) A change in an analytical procedure used for testing components, packaging components, the final intermediate, in-process materials after the final intermediate, or starting materials introduced after the final intermediate that provides the same or increased assurance of the identity, strength, quality, purity, or potency of the material being tested as the analytical procedure described in the approved application.

2.2.3.4 Minor Changes – Annual Report

The followings are examples of changes in specifications considered to have a minimal potential to have an adverse effect on the identity, strength, quality, purity, or potency of a drug product as these factors may relate to the safety or effectiveness of the drug product.

1) Any change in a specification made to comply with an official compendium, except the changes made to relax an acceptance criterion or delete a test to comply with an official compendium that is consistent with FDA statutory and regulatory requirements, for which a CBE-30 is required.
2) For drug substance and drug product, the addition or revision of an alternative analytical procedure that provides the same or increased assurance of the identity, strength, quality, purity, or potency of the material being tested as the analytical procedure described in the approved application or deletion of an alternative analytical procedure.
3) Tightening of acceptance criteria.
4) A change in an analytical procedure used for testing raw materials used in drug substance synthesis, starting materials introduced prior to the final drug substance intermediate, in-process materials prior to the final intermediate, or drug substance intermediates (excluding final intermediate) that provides the same or increased assurance of the identity, strength, quality, purity, or potency of the material being tested as the analytical procedure described in the approved application.

2.2.4 Analytical Development with FDA Guidelines in Mind

Knowing and understanding the above FDA guidelines is the minimum requirement for a lead analytical scientist. Among the critical quality attributes of an analytical method, meeting regulatory requirements is the No. 1 priority. That is not just because those are the laws and guidelines that we have to follow; it is also because the analytical development strategy and approaches are determined based on the regulatory filing strategies. "Begin with an end in mind" is one of the seven habits of highly effective people [6]. A clear understanding of the CMC contents, requirements, and the company's filing strategy of a new product can guide the lead analytical scientist to plan analytical activities accordingly and spend the right amount of energy and time on method development, method validation, and product evaluation (testing).

For example, in the case of an NDA product development, even if the API may be a well-known molecule, such as naproxen sodium, loratadine, cetirizine HCl, ibuprofen, etc., the analytical methods must be developed with thorough evaluations from all angles, including accuracy in potency analysis, stability-indicating power, and sensitivity of the degradation method, discriminative power of the dissolution method, photosensitivity, etc. The method must be robust and rugged so it can be used routinely and can be transferred to QC laboratories that may be located globally. Moreover, before product registration, clinical trials are usually associated with an NDA filing strategy. Although the requirements for the analytical methods used to support the release and stability testing of the clinical trial materials or clinical supplies are clinical study phase-dependent, the principle is that the method must be evaluated based on common validation parameters such as linearity, specificity, accuracy, precision, sensitivity, etc. [7] Therefore, the lead analytical scientist should plan well to make sure the method is "validatable." Although at the clinical trial stage, it is not necessary to have a formal method validation report, a certain level of semi-formal documentation, which summarizes the outcome of method evaluation, is highly desired before using the method to conduct the release testing of the clinical materials. From a project management point of view, this means the method development timeline has to align well with that of formulation development, scale-up development, and clinical study arrangement.

In the case of an ANDA, the method must be able to show equivalency between the product under development and the reference product. Many pharmaceutical companies who make generic drugs are good at reverse engineering, or deformulation [8]. Advanced analytical instrumentation such as LC-MS, GC-MS, NMR, FT-IR, ICP, XRD, etc. are employed to separate the excipients from the API(s) and analyze the degradation profiles of the product. A crucial condition for a generic drug product to get approval from the FDA is the demonstration of the dissolution profile similarity between the generic product and the reference product. For analytical scientists who work in those areas, the ability to quickly develop suitable dissolution methods is a must to help the company to use the dissolution profiles as the surrogate for demonstrating bioequivalence.

If the application is an sNDA, then the flexibility in method development might be limited. Since it is a supplement, an alteration, to an approved NDA, as described before, depending on the changes, the filing strategies are different. For example, it usually would not be the desired path to widen the specification limits for an sNDA. Therefore, if during the product development, there are new impurities or different dissolution behaviors observed, the analytical scientist should bring attention to the team to get alignment on the strategy of the path forward. The analytical development also needs to provide opinions/suggestions to the team when deciding whether a change in analytical method is Major or Minor, whether it can be filed as a CBE or CBE-30, or can be just included in an annual report.

2.3 ICH Guidelines and Analytical Development

In addition to the FDA, several other institutes also publish guidelines of which an analytical scientist should be aware. Among them, ICH issues a number of guidelines in four major categories: Quality, Safety, Efficacy, and Multidisciplinary. The guidelines in the Quality category directly govern the analytical development work. Many ICH guidelines are very famous. Every analytical scientist who works in the pharmaceutical industry should study at least some, if not all, of the ICH guidelines during their work. Listing detailed descriptions of the guidelines in this book will not be necessary. A high-level description of the ICH guidelines in the Quality category is summarized in Table 2.1.

The ICH guidelines should be treated as practical guidance for the daily work of analytical development. As an example, we can take a look at the ICH Q3B: "IMPURITIES IN NEW DRUG PRODUCTS" [9]. First of all, an analytical development scientist should be familiar with the terminology of impurities and degradation products. The degradation products with specific acceptance criteria included in the specification for a new drug product are referred to as *specified* degradation products. Within this *specified* category, there are two subcategories which are defined as (1) *Specified Identified* degradation product; (2) *Specified Unidentified* degradation product. *Identified* here means the molecular structure of the compound is identified and confirmed. The level of the specification limit for that compound is determined based on the toxicity studies or based on other scientific justifications. The specification usually spells out the compound's chemical name or an abbreviation of the chemical name. Some naming conventions such as Degradation Product A, B, C, D should be used with caution, although it is commonly seen and seems convenient to the scientists who are using those terms for the projects at hand. The issue may arise later when the information needs to be transferred to someone who has no prior knowledge of the product or those degradation compounds. Moreover, some pharmacopeias also use related compound or degradation product A, B, C, D to name the degradation compounds, and it can be very confusing to remember that the USP A is the EP B, or the USP C corresponds to the JP F, etc. In some extreme cases, even a certificate of analysis only lists the compound name as "XYZ (Name of the API) related compound H" or "ABC12345" without any reference to the real chemical name of the compound. The manufacturing QC laboratory tries to use the method and then finds out no one knows what that compound really is. The word *Unidentified* here means the molecular structure of a compound is unknown,

Table 2.1 ICH guidelines regarding quality of drug substances and products.

Document number	Guideline content
Q1A–Q1E	Stability Testing; Photostability Testing; Bracketing and Matrixing Designs for Stability Testing of New Drug Substances and Products; New Dosage Forms; Evaluation of Stability Data
Q2	Analytical Validation
Q3A–Q3D	Impurities in New Drug Substances and Products; Guideline for Residual Solvents; Guideline for Elemental Impurities, etc.
Q4–Q4B	Pharmacopoeias; Pharmacopoeial Harmonization; Microbiological Examination of Non-Sterile Products; Disintegration Test; Uniformity of Dosage Units; Dissolution Test; Sterility Test; Tablet Friability; Polyacrylamide Gel Electrophoresis; Capillary Electrophoresis; Analytical Sieving; Bulk Density and Tapped Density of Powders; Bacterial Endotoxins Test; etc.
Q5A–Q5E	Quality of Biotechnological Products
Q6A–Q6B	Q6A Specifications: Test Procedures and Acceptance Criteria for New Drug Substances and New Drug Products: Chemical Substances; Q6B Specifications: Test Procedures and Acceptance Criteria for Biotechnological/Biological Products
Q7	Good Manufacturing Practice
Q8	Pharmaceutical Development
Q9	Quality Risk Management
Q10	Pharmaceutical Quality System
Q11	Development and Manufacture of Drug Substances
Q12	Technical and Regulatory Considerations for Pharmaceutical Product Lifecycle Management
Q13	Continuous Manufacturing of Drug Substance and Drug Products
Q14	Analytical Procedure Development

regardless of exhaustive endeavors for molecular structure elucidation. Many analytical procedures use relative retention time (RRT), such as "RRT 0.38" to name such a *Specified Unidentified* peak in the specification. This is not necessarily a good practice since the RRTs can be different when obtained at different laboratories or under slightly different chromatographic conditions, and thus may confuse peak identification during routine analysis. Instead, those peaks can be named in a different way, such as "XYZ (the name of the API) related specified unidentified compound I," "XYZ (the name of the API) related specified unidentified compound II," and the RRTs are provided side by side, but only as a reference.

Besides the identified degradation products, there is a third category, that is, the *Unspecified* degradation products. Those peaks belong to a category that they are either not expected (i.e. truly unknown to the method developer and thus *Unspecified*) or they are known to exist but at levels that are much lower than the reporting threshold defined by the ICH guideline, and therefore can be treated as *Unspecified*. Note the terminology of *unknown* is not officially seen in the guidelines, and should be used with caution.

Another critical guidance provided by the ICH Q3B is the reporting–identification–qualification thresholds allowed for impurities in drug products [9]. *Qualification* is the process of acquiring and evaluating data that establishes the biological safety of an individual degradation product or a given degradation profile. It is required to provide rationales, including safety considerations, for establishing acceptance criteria that are higher than the corresponding ICH qualification threshold for degradation products.

Understanding and *remembering* the reporting–identification–qualification thresholds of impurities in drug products are the basis for setting up appropriate product specifications. The FDA requires that a product specification must include the specified identified degradation products together with the specified unidentified degradation products that are estimated to be present at a level greater than the identification threshold given in ICH Q3B. For degradation products known to be unusually potent or to produce toxic or unexpected pharmacological effects, the FDA [10] requires that the quantitation and/or detection limit of the analytical procedures correspond to the level at which the degradation products are expected to be controlled. For unidentified degradation products to be listed on the drug product specification, the FDA requires clear descriptions regarding the procedure used and assumptions made in establishing the level of the degradation product. The acceptance criteria for those unidentified degradation products should not be more than the identification threshold defined in Q3B. In addition, when establishing degradation product acceptance criteria, the first critical consideration is whether a degradation product is specified in the USP. If there is a monograph in the USP that includes a limit for a specified identified degradation product, the acceptance criterion should not be set higher than the USP limit [11]. If the level of the degradation product is above the level specified in the USP, qualification work to demonstrate the safety of the compound and to demonstrate the suitability of the proposed level is needed. Then, with appropriate toxicity qualifications, petitioning the USP for revision of the degradation product's acceptance criterion is needed. If the acceptance criterion for a specified degradation product does not exist in the USP and this degradation product can be qualified by comparison to an RLD, the acceptance criterion should be similar to the level observed in the RLD. In other circumstances, the acceptance criterion may need to be set lower than the qualified level to ensure drug product quality. Some special cases exist.

For example, a specified identified degradation product is considered qualified when it meets one or more of the following conditions:

1) When the observed level and proposed acceptance criterion for the degradation product do not exceed the level observed in the RLD.
2) When the degradation product is a significant metabolite of the drug substance.
3) When the observed level and the proposed acceptance criterion for the degradation product are adequately justified by the scientific literature.
4) When the observed level and proposed acceptance criterion for the degradation product do not exceed the level that has been adequately evaluated in toxicology studies.

Note that although Quantitative Structure–Activity Relationship programs may be used for prediction of the toxicity of an individual degradation product or a given degradation profile, the results are not generally considered conclusive for qualification purposes [11].

Similar to the Q3B, ICH Q3A [12] governs the controls of impurities in drug substances for NDAs, which also applies to ANDAs. Where applicable, the drug substance specification should include a list of the following types of impurities:

1) Organic impurities that include: Each specified identified impurity – Each specified unidentified impurity – Any unspecified impurity with an acceptance criterion of not more than ($\leq$) the identification threshold in Q3A(R) – Total impurities
2) Residual solvents
3) Inorganic impurities

When establishing impurity acceptance criteria for the drug substance, the first critical consideration is whether an impurity is specified in the USP. If there is a monograph in the USP that includes a limit for a specified impurity, the acceptance criterion should not be higher than the official compendial limit [13].

All the above information is important and is straightforwardly from the guidelines. What is more important is how we use that information. As one example, the ICH Q3B can serve as a basis for an analytical scientist to establish the analysis range of a method, as described below.

At the beginning of method development, the first question an analytical scientist should ask the project team (especially to the medical affairs colleagues) is "what is the proposed maximum daily dose of the API?" With that information, the corresponding reporting threshold of the degradation product can be determined based on the levels defined in the ICH Q3B. For example, if the maximum daily dose of an API is 10 mg, then the reporting threshold of degradation products is 0.1% (weight/weight, mg/mg) of the product label claim. To ensure sufficient sensitivity of the method, the target method Limit of Quantitation (LOQ) will be set at half of that reporting threshold, i.e. at 0.05% of the API label claim. With that goal in mind, once a method with sufficient separation power has been developed, the analytical scientist can conduct the following experiments to determine the analytical sample concentration range:

1) Select an appropriate composition of the diluent based on the solubility of the API and its related compounds. Pay attention to the potential impact on the chromatography from the solvent and additives used in the diluent (such as diluent-mobile phase mismatch in terms of solvent strength).
2) Pick the related compound that has the weakest UV absorbance at the selected monitoring wavelength.
3) If no related compound is available (especially at the beginning of the method development), use API itself as a surrogate.
4) Prepare a sample solution by dissolving 5, or 10, or 20 units of the product into, for example, 250 mL of diluent to make a "100% level" solution. This "100% level" is tentative at this moment.
5) Dilute the "100% level" solution 2000 times to obtain a 0.05% solution, label as "LOQ Trial 1 Solution."
6) Inject the "LOQ Trial 1 Solution" onto HPLC with injection volumes from 10 to 100 µl.
7) Evaluate the observed signal-to-noise (S/N) ratios. Although an S/N ratio of 10 is acceptable for the LOQ level, to ensure the robustness and ruggedness of the method, an S/N ratio of more than 20–30 should be the target for setting up the method LOQ. The injection volume should be selected based on the S/N ratio and chromatographic baseline smoothness and peak shape.
8) If the S/N is more than 20–30 with a reasonable injection volume then the above chromatographic parameters, i.e. the composition of the diluent, the "100% level" solution concentration, and the injection volume, are settled.
9) If the S/N is less than 20 or a nonoptimum chromatography is obtained, the parameters can be adjusted until an acceptable combination of the above parameters are achieved.

The above steps illustrate a workflow for establishing the lowest level that the method sensitivity must achieve for degradation product quantitation purposes. As a side note, when determining the 100% analytical concentration level for *Assay* analysis, on the other hand, the UV absorbance intensity of the 100% level injection should be at around 0.5–0.8 AU to ensure a linear response can be achieved should the upper limit of the assay analysis range is 150%. The solution prepared in Step 4 should then be diluted accordingly.

One huge topic in the ICH guidelines that a lead analytical scientist should know, but this book will skip, is the stability study. The reasons to skip this topic are (1) the topic is so big that itself can be the focus of a book; (2) there are numerous references, literature, training courses available for interested readers to get self-educated on this topic.

2.4 Pharmacopoeia Monographs and Analytical Development

Every country has its own pharmacopeia and requirements on analytical testing of excipients, APIs, and/or drug products. In this book, only the USP–NF system will be briefly described. USP-NF is a combination of two compendia, the United States Pharmacopeia (USP) and the National Formulary (NF) [14]. Monographs for drug substances, dosage forms, and compounded preparations are in the USP. Monographs for dietary supplements and ingredients appear in a separate section of the USP. Excipient monographs are in the NF. A monograph includes the name of the ingredient or preparation; the definition; packaging, storage, and labeling requirements; and the specification. The specification consists of a series of tests, procedures for the tests, and acceptance criteria. These tests and procedures require the use of official USP Reference Standards, a requirement of which may not be known to many analytical scientists. USP was originally founded in 1820 by a group of physicians who wanted to standardize the preparation and use of medicinal products. Today, USP is a non-profit scientific organization whose mission is to improve public health around the globe. USP develops public standards for identity, strength, quality, and purity of medicines, food ingredients, and supplements. Those public Standards are described in a compendium of monographs and general chapters. The USP monographs are considered as official testing procedures. A drug product is regarded as (1) adulterated if the drug is represented as one in the USP, and its strength, quality, and purity differ from the standards in the USP or (2) misbranded if a product does not meet standards for packaging and labeling in the USP. However, the legal enforcement of USP standards is carried out by the FDA, and not the USP. It is acceptable to use alternative analytical procedures as long as a full method validation report is available.

In addition to those individual USP monographs, USP has many general chapters that describe requirements for conducting analytical laboratory work and for conducting data analysis. Chapters numbered less than <1000> are enforceable, and chapters numbered greater than <1000> are for information. There are many general chapters that are applicable to the analytical development, among them a few examples are listed here, such as General Chapter <621> Chromatography, General Chapter <701> Disintegration, General Chapter <711> Dissolution, General Chapter <791> pH, General Chapter <905> Uniformity of Dosage Units, General Chapter <1010> Analytical Data – Interpretation and Treatment, General Chapter <1210> Statistical Tools for Procedure Validation, General Chapter <1220> The Analytical Procedure Lifecycle, General Chapter <1251> Weighing on an Analytical Balance.

Some (by no means inclusive) other special general chapters are listed here for information:

<231>   Heavy Metals
<232>   Elemental Impurities – Limits

<233>   Elemental Impurities – Procedures

<467>   Residual Solvents/Organic Volatile Impurities

<61>   Microbiological Examination of Nonsterile Products: Microbial Enumeration Tests

<62>   Microbiological Examination of Nonsterile Products: Tests for Specified Microorganisms

<1111>   Microbiological Examination of Nonsterile Products: Acceptance Criteria for Pharmaceutical Preparations and Substances for Pharmaceutical Use

As mentioned before, a lead analytical scientist should have the knowledge of the USP requirements. At the beginning of the project, depending on the availability of a USP monograph method, the analytical method development strategy varies. Although in many cases, the USP monograph methodology needs to be updated to keep up with the fast-growing technologies, the USP methods are still considered as the public standards. In-house developed methods must show equivalency or can be demonstrated a better performance than the USP methods. Equally, if not more important requirement for a lead analytical scientist is that he/she should study, understand, and memorize the key points in those USP general chapters. The contents in the general chapters are directional. Without knowing that, an analytical development strategy could potentially be off-track. In addition, careful examination of the difference between USP and other pharmacopeias such as EP or BP is needed if the product is developed with countries outside the United States in scope. For example, the USP general chapter <701> Disintegration does not specify the temperatures when conducting disintegration experiments for effervescent tablets, which reads: "Place 1 tablet in each of 6 beakers containing 200 mL of water. A suitable beaker will have a nominal volume of 250–400 mL" [15]; while the European Pharmacopoeia specifies the use of six tablets, one at a time, to be tested in a beaker containing 200 mL of water at 15–25°C [16].

2.5 Formulation Development and Analytical Development

Formulation development starts from, although not necessarily be always the case, the so-called preformulation, followed by the formulation and scale-up, continues to technology transfer, and the final manufacturing process validation.

Preformulation is to characterize the drug molecule(s) for chemical and physicochemical properties such as molecular structure, ionization (pK_a), hydrophobicity (log P), stability, polymorphism, solubility, moisture sorption, surface morphology, surface area, melting point, particle size distribution, etc. [17–20] At the preformulation stage, the formulation scientists are mainly concerned about the question of "Go" or "No Go." The purpose can be formula feasibility scouting or to quickly identify potential formulation failures. The prototypes are usually made at small laboratory bench scales on miniature equipment. At this stage, formulation scientists carry out evaluations on excipient compatibility, powder flowability, tablet compressibility, organoleptic properties, and of course, product physical and chemical stabilities. Expectations on the analytical scientists at the preformulation stage are timely delivery of data. At the same time, though almost contradictory to the desire for speedy delivery, the expectations also include insightful explanations of the analytical results. The (unconscious or unsaid) assumption behind those expectations is that the analytical scientists already have (appropriate) method(s) for analyzing the prototypes, and the analytical scientists know what the data are revealing. However, quite oppositely, at the time of initiating new product development, no one on the team really has all the knowledge or the crystal ball to know what may be happening. It is possible that the formulation scientists are making the prototypes the first time, and the analytical scientists may have no or limited relevant experiences with the new formulas from the analytical methodology point of view.

Once promising prototypes are identified, the formulation development starts to focus on the evaluation of (1) the long-term stability of the formula(s) from chemical stability and physical performance perspectives, (2) the packaging variations to seek most effective and economically efficient packaging configurations in addition to appealing designs, and (3) the feasibility to scale up the formulation process from the laboratory scales to pilot scales (usually is 1/10 of the final commercial scales) and eventually to the commercial manufacturing scales. During the laboratory formulation development, the formulation scientists need timely but accurate and precise analytical results to help identify the manufacturing parameters for establishing a designed space of operation.

Whether it is preformulation or formulation, whether it is desired to have accurate analytical data or rough outcomes, it seems that a quick turnaround of the analytical results is a general request. This is, however, contradictory to the nature of analytical work and also is contradictory to the nature of many analytical scientists. In general, analytical scientists are detail-oriented, and who may think: "Developing a perfect method is my job." Analytical method development such as chromatographic procedure development is a time-consuming journey that requires thorough scrutiny of the separation and quantitation of chromatographic peaks. The methods have to be robust so that they can be transferred to and are used routinely in quality control laboratories. Indeed, many Quality by Design (QbD) or Design of Experiments approaches found in the literature regarding method development often describe comprehensive method development workflows that seem not to be constrained by timelines or resources.

2.5.1 Method Development Based on an Ideal, Comprehensive Quality by Design

QbD has been the topic of many regulatory guidelines, such as ICH Q8 – Pharmaceutical Development; ICH Q9 – Quality Risk Management; ICH Q10 – Pharmaceutical Quality System; and ICH Q11 – Development and Manufacture of Drug Substances. QbD in general refers to a systematic approach that begins with predefined objectives and emphasizes product and process understanding and process control, based on sound science and quality risk management. The goal is to embed quality into pharmaceutical products to ultimately protect patient safety. The same concept applies to analytical methods development. Some key points in methods development QbD are listed below:

1) Analytical Target Profile (ATP) – As mentioned previously, the concept of an ATP parallels the concept of a QTPP described and defined in ICH Q8. The ATP defines the objectives of the test and quality requirements, which include the expected level of confidence for the reportable result, and allows the method developer to draw correct conclusions regarding the measured attributes of the material.

2) Risk Management – Quality Risk Management (QRM) for analytical procedures is a systematic process for the assessment, control, communication, and review of risks to the quality of data across the product life cycle. Process mapping, Ishikawa (fishbone) diagrams, etc. ensure a rigorous approach in identifying all potential variables that may affect data quality. The variables should include all aspects of the full analytical procedure, i.e. sampling, sample preparation, standards, reagents, facility, and equipment operating conditions.

3) Knowledge Management – The method-developing laboratory should store the obtained knowledge gathered in a central location and makes the information easily sharable among different laboratories.

4) Analytical Control Strategy – The procedures should explicitly specify the variables and their acceptable ranges (from the risk assessment or experimental work).

Some template wordings of an ATP can be found in the USP stimuli paper for its general chapter <1220> "The Analytical Procedure Lifecycle." For example, an ATP can be expressed as: "*The procedure must be able to quantify [analyte] in the [description of test article] in the presence of [x, y, z] with the following requirements for the reportable values: Accuracy = $100\% \pm D\%$ and Precision $\leq E\%$.*" The ATP inputs for [analyte], [description of test article] and [x, y, z] (which may be impurities or excipients) can be specified. Values for D and E should be specified. For example, D represents the acceptable variation range and E represents desired upper limit of the relative standard deviation (%RSD). Another example wording reads: "*The procedure must be able to quantify [analyte] in the [description of test article] in the presence of [x, y, z] so that the reportable values fall within a TMU of ±C%.*" This example contains criteria for the Target Measurement Uncertainty (TMU) of ±*C%*, which is directly linked to the results generated by the procedure. The TMU considers the acceptable difference between the measured reportable value and the target value. TMU can be established based on a fraction of the specification range. For example, if the specification range is ±10%, a TMU can be set at ±5%. The following are described in the same stimuli paper:

> When establishing an ATP, the following should be considered, where relevant:
> - Sample to be tested
> - Matrix in which the analyte will be present
> - Allowable error for the measurement as assessed through accuracy (bias) and precision, both of which make up the TMU
> - Allowable risk of the criteria not being met (proportion of results that are expected to be within the acceptance criteria)
> - Assurance that the measurement uncertainty and risk criteria are met
>   ...
> This approach encourages understanding and control of sources of variability ...

However, although the literature discussions on analytical QbD or the above USP examples of ATP make it look like the analytical QbD is only about a thorough understanding of one method, in the fast-paced pharmaceutical product development environment, we have to think the QbD beyond just developing one perfect method. Indeed, in order to achieve the desired ATP, lots of work must be carried out. Figures 2.3 through 2.7 depict how one chromatographic method can be developed in an ideal world if the analytical scientists have sufficient time and resources. The figures include what is needed from knowledge perspectives, from material readiness perspectives, and from time perspectives. The graphs are only for illustration purposes, and by no means refer to any real practices.

In Figure 2.3, the method development starts with a competent analytical scientist who has the knowledge of not only separation science but also organic chemistry. The chemistry knowledge enables the scientist to thoroughly study the molecular structures. He/she can predict the molecular interactions among the analytes; between the analytes and the chromatography column surface stationary phases; and among the analytes, column surfaces, and the mobile phase components. The chemistry knowledge also enables the scientist to anticipate the formation of potential degradation products of the API and possible reaction

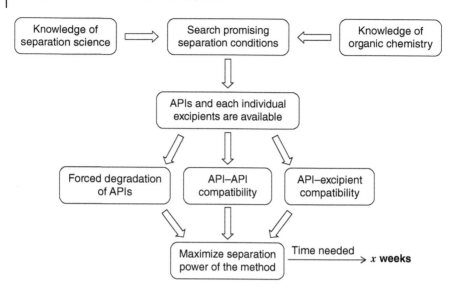

Figure 2.3 Ideal workflow of chromatographic method development for degradation product analysis.

products formed between the API and the excipients in the drug product. The knowledge of separation science enables the scientist to understand the features of different column surface chemistry, to be familiar with the characteristics of the organic solvents commonly used in mobile phases, etc. With both separation science and organic chemistry knowledge in mind, the scientist can start to find promising separation conditions, by either a thorough literature search, or by starting trials based on educated guesses. From the material point of view, to do a thorough method development, the analytical scientist needs to have the API and all the excipients. That seems to be a no-brainer requirement in a QbD context. It is, however, not necessarily an easy task to get all the required materials at the beginning of a new product development. In many cases, the excipients are not available or are even not decided by the formulation scientists since product development is still at its initiation stage. High flexibility to cope with various situations and the ability to make decisions to move forward without all the information available are required skills for a lead analytical scientist to have. Equipped with the chemistry knowledge and with (at least some) materials in hand, the analytical scientists start to work on forced degradation studies of the API(s) to understand the intrinsic chemistry of the molecule. The samples from the forced degradation studies, oftentimes, although not all the times, can serve as the worst-case scenarios to challenge the method separation power. In the meantime, the analytical scientist starts to evaluate the chemical compatibility between the API and excipients, and sometimes between the APIs if the product contains multiple APIs. This step serves two purposes, one is to warn the formulation scientist potential incompatibility between the API and excipients, and the other one is to further challenge the separation power of the method, and to help the analytical scientist distinguish the excipient related peaks from the API related peaks. The workload of the stress studies and the compatibility studies can be enormous, and therefore the thus developed method may have maximized separation power. The by-product of that approach is, however, a long leading time needed to complete such a massive work.

Finding out chromatographic conditions is, however, only half of the story. The other half of the method development is sample preparation. As depicted in Figure 2.4, the method development for sample preparation also starts with a competent analytical scientist who has ample knowledge of analytical sample preparation techniques and who is familiar with material properties from both chemical and physical aspects. With that knowledge in mind, the analytical scientist can start working on developing robust sample preparation procedures. Again, the availability of the products or formulation prototypes is imperative. It seems to be a *chicken-or-egg* situation. At the early stage of the formulation development, the prototypes are not produced at an amount that enables the analytical scientists to freely explore the best sample preparation approaches. However, without a sufficient amount of samples to facilitate the analytical scientists to develop robust sample preparation procedures, the formulation scientists may not be able to get the much-needed accurate data. The analytical scientists have to evaluate the extraction effectiveness and efficiency not only from the freshly made prototypes but also from those stressed prototypes to mimic future sample preparations for products that will be stored under high

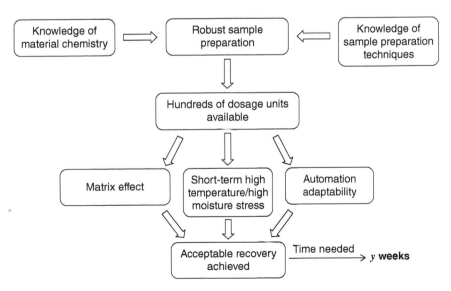

Figure 2.4 Ideal workflow for sample preparation development.

temperatures and high humidity for a long time. The analytical scientists also need to eliminate sample matrix effects in order to achieve accurate results [21]. In some cases, to gain a high throughput from a QC laboratory, automated sample preparation is desired. That can put an extra burden on sample preparation development, and as a consequence, more time is needed.

Up to this point, a method for the analysis of API-related compounds, including degradation products and impurities, has potentially been developed. With some tweaks, an assay and a content uniformity method can be developed based on the related compound analysis method, without having to spend too much time. On the other hand, although we can assume that the chromatographic conditions suitable for the assay analysis can be used for analyzing dissolution samples, that is, however, again, just half of the story. The other half of the dissolution method development is actually more time consuming, which is, the development of a meaningful dissolution condition.

Finding a dissolution condition that is discriminatory and meaningful is very time-consuming (Figure 2.5). Most countries consider dissolution methods are for quality control purposes. Some health authorities may emphasize achieving in-vitro in-vivo correlation. Regardless, dissolution profiles should be obtained during the formulation development [22]. To achieve the so-called discriminatory power of the method, various parameters have to be evaluated. Among them are the changes in API properties such as particle size or particle size distribution, significant changes in formula, different shear forces applied during wet granulations, different pressures applied during the compression, etc. In addition to the

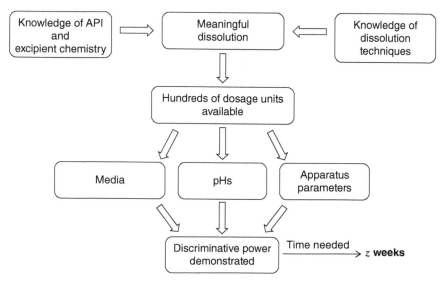

Figure 2.5 Ideal workflow for dissolution method development.

changes in raw materials, formulations, or manufacturing processes, aging of the products or different storage conditions can also bring variations in the dissolution performance. Oftentimes, the analytical scientists develop a dissolution method using freshly made prototypes at laboratory scales. The dissolution can slow down dramatically after the prototypes are put on stability for a certain period, especially for gelatin related dosage forms such as hard shell capsules. It is a well-known fact that cross-linking (polymerization) of the gelatin shell materials can slow down the disintegration of the formulation. The cross-linking process is facilitated by high temperature, high humidity, photostress, or the existence of aldehydes or peroxides [23]. Although the addition of digestive enzymes such as pepsin or pancreatin is recommended by USP <711> to help recover the dissolution rate, this approach does not always work. In some solid dosage forms such as orally disintegrating tablet, the so-called superdisintegrants such as crospovidone, croscarmellose sodium, and sodium starch glycolate, are added. Those water-craving ingredients are incorporated into the formula to facilitate the disintegration of the tablets when in touch with the dissolution media. However, if the packaging is not protective enough, the superdisintegrants absorb water/moisture gradually during the storage. The end result of that gradual water absorption is that after the tablets have been stored three or six months under 40°C/75%RH (Relative Humidity), and then are put into the dissolution media, the superdisintegrants do not crave for water as much as they are supposed to, and thus the disintegration of the tablets is not as accelerated as the formulation is designed. The dissolution method, therefore, should be able to differentiate dissolution performances between freshly made capsules or tablets and those capsules or tablets that are stored under different storage conditions as well. Also, different pHs of the dissolution media may need to be evaluated, especially for bioequivalence evaluation. At a minimum, dissolution profiles obtained at pH 1.2, 4.5, and 6.8 should be evaluated [24, 25]. Since vastly different solubility of the API at different pHs may exist, more challenges in the method development can be expected. On top of the discriminatory requirement, a dissolution method must be meaningful. The dissolution rate can be made fast or slow, depending on the selected dissolution media pH, the addition of some additives such as surfactants and even organic solvents, and the choice of apparatus, sinkers, and rotation speeds. A sound scientific justification should exist before finalizing the dissolution method development. As can be imagined, to cover all the above aspects, a large amount of time is needed for such endeavors.

After weeks or months of hard work, and with detail-oriented attitude of the lead analytical scientist, the methods for assay, dissolution, and related compound analysis are finally developed. The methods are then used for testing developmental long-term stability samples and are further scrutinized to see if there are places to improve (Figure 2.6). Usually, a three-month stability study is adequate for the analytical scientist to evaluate the method performance and readiness. Once the

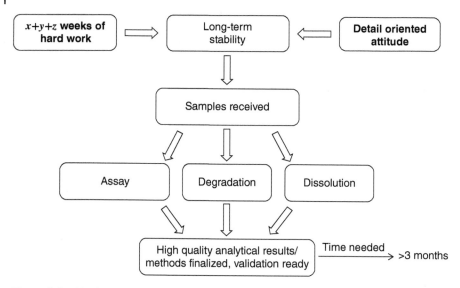

Figure 2.6 Ideal workflow for method finalization.

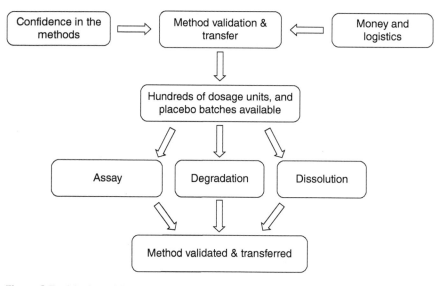

Figure 2.7 Ideal workflow for method validation and transfer.

method has withstood the tests, the analytical scientists will move on to method validation and transfer.

Method validation and transfer are the final chapters of the method development (Figure 2.7). The analytical scientists will summon all the remaining energy to perform some prevalidation work to double-check the key parameters of the method, gain much-needed confidence in the soon-to-come method validation/transfer, prepare a detailed method validation/transfer protocol, and finally, carry out the validation work. They can finally take a breath of relief after a successful validation and transfer. As shown in Figure 2.7, money and logistics come into the picture. The team may have to spend a fortune on reference standards and have to coordinate the shipping of samples cross the oceans, if global manufacturing sites are involved.

From Figures 2.3 through 2.7, we can see that adequate time and resources are fundamental requirements to go through such a comprehensive QbD path of method development.

Unfortunately, the gap between the expectations from the formulation scientists for quick and accurate analyses plus insightful explanations of the results, and the efforts and resources needed for analytical method development is not ignorable. As Figures 2.3 through 2.7 illustrate, the actual work does not start until the analytical scientists physically possess the necessary materials (samples, placebos, excipients, etc.) and have a clear vision of what the methods are trying to achieve. The knowledge of molecular structures (organic chemistry), sample preparation and matrix effects (analytical chemistry, physical chemistry, and material sciences), measurement techniques (instrument analysis), and measurement uncertainties (statistics) becomes a necessity in order to design the quality attributes into the methods. In reality, however, it is not rare to see some analytical scientists implementing the QbD by checking the boxes on some general QbD list or making some drawing based on more or less universal fishbone diagrams that contain the same critical quality attributes regardless of what methods they are going to develop. Moreover, in the fast-paced product development environment, formulation and analytical development can occur in parallel. All that the analytical scientists have at the early stage of product development is a limited amount of prototypes. In addition, the product development in the OTC industry faces frequent formulation changes that are not initiated by the formulation scientists but are directed by consumer desires, marketing strategies, regulatory climate changes, packaging and manufacturing capabilities or restrictions, etc. Formulation or manufacturing processes can change at the last minute due to a more desired element is proposed to the formulation development, such as a different color, a different flavor, a cheaper ingredient, or a more capital friendly manufacturing process. In an ideal world, analytical procedures are formulation specific. Different formulas can require different analytical methods. Although nowhere one can find a statement such as comprehensive QbD approaches should mainly apply to a method that is developed for a product that has a long-term life cycle, and/or the product is so unique that there is no similar analytical method available, it is most likely in reality that one cannot apply such comprehensive

QbD approaches presented in Figures 2.3 through 2.7 straightforwardly in the analytical development work for fast-paced product development. In some extreme cases, the $(x + y + z)$ weeks or months requested by the analytical development team would make the formulation colleagues confused and would even cause frustrations among the team members.

The core value of a QbD is to build quality into the methodology. It is not a show-off of how comprehensive an approach one has taken. To make the concept of QbD relevant and meaningful to the method development in fast-paced pharmaceutical product development, fit-for-purpose, teamwork, knowledge sharing, and platform approach are some key elements to success. We need to think QbD beyond just developing ONE perfect method by one analytical scientist.

2.5.2 Fit-for-Purpose, Teamwork, Knowledge Sharing, and Platform Approach

Ideally, a stability-indicating HPLC method with powerful separation capability should be developed to provide ample stability information on the API, the formula, and the final packaged product. Such a method has to be developed based on molecular structure examination, forced degradation, and drug-excipient compatibility studies. This method shall properly retain as many as possible the compounds that form as a result of API degradation or its reaction with other compounds in the formula. The method has to be able to retain a wide range of molecules with drastically different polarity and hydrophobicity. The analytical scientist should make efforts to study the properties of excipients in addition to studying the properties of APIs. Each inactive ingredient in the formula serves different purposes, and as a whole, they ensure the product performs as designed. Even though the excipients are regarded as inert, they are chemicals that possess functional groups. Moreover, some residual impurities left in the excipients during the synthesis of those materials can cause API degradation. However, developing/optimizing/fine-tuning an analytical method in parallel with the use of that method which is still under development to perform prototype testing can be resource-demanding and causes delays. What might happen include (1) analytical scientists "hold" the data/results without timely sharing them with the team; (2) analytical scientists get self-confused about the peak tracking and report self-conflicting stability testing results; and (3) resulting in lower level of trust among team members. All those unproductive happenings have the same underlying root cause, that is, the lack of confidence in the method that is being used. The analytical scientists are not sure about the method performance and are trying to double-check, and double-check, before sharing the results with the team. From the eyes of other team members, the analytical development is holding the results and is not transparent. On the other hand, since the method is being constantly tweaked, after a while, the method can have very different peak retention profiles. When the analytical scientists are in a time crunch to present the results to the team, they may not realize the peak of interest has shifted its retention time, and another peak is misplaced as the peak of interest. Ironically, this type of error can often be caught by other project team members during the presentation. To the analytical scientist who has worked so hard, it becomes an embarrassing moment. When this kind of event happens a few times, the trust level among team members inevitably gets lower.

To conduct high-quality analytical work within the shortest possible time, the first-thing-first is for the analytical scientists to have a mindset of fit-for-purpose. Only after a clear purpose is defined, the analytical scientists can determine whether the approaches taken are fit or not. From the big-picture point of view, the overall purpose is to gain a balance between developing a perfect analytical method and concurrently providing adequate support to formulation development. In other words, fit-for-purpose in this context regards more about taking product development stage-appropriate approaches and having flexible mindsets. For example, the purpose of an analytical method for preformulation should be that the method can provide fast results, and provide directional guidance. The method then needs to have some character traits such as short run time, easy-to-prepare mobile phases which have long shelf-life (at least a few weeks to a few months at ambient laboratory conditions). The method does not have to possess a separation power to cover all possible related compounds. Somewhat counter-intuitively, the analytical scientists also need to have the ability to tolerate the imperfection of the method and let-go of the desire of pursuing detailed method fine-tuning at this stage. Once the development moves on from preformulation to formulation development, the analytical scientists will need to conduct analyses, obtain results, and provide as much insight from the testing as possible within a time as short as possible. At this stage, the separation power of the method becomes desired and the accuracy of the method becomes a must. The fit-for-purpose then contains a different notion. The purpose is not only fast but also accurate and comprehensive. As we have already seen, for one analytical scientist to get all the method development job done by taking some, if not all, QbD desired approaches, the mission is very tough, to say the least. It seems that the conflict between detail-oriented analytical method development and fast-moving formulation and process development is about to happen.

To overcome the hurdles, teamwork and knowledge sharing are the keys to success. Nowadays, company organization setups often require more work done with fewer resources. However, lowering the quality of our work is not an option.

Spending that much time to develop one perfect method by one analytical scientist in parallel with formulation development is mission impossible, too. The only way out is to rely on the team effort. We should use the whole analytical development team to develop a few powerful methods by letting different team members work on different portions of the method development and encourage knowledge sharing among the team. It is definitely not an affordable approach to cultivate silos at work.

The analytical development as a team should work together and help each other, *spontaneously*, by the design of workflow or process, and not by management push. The concept of *spontaneously* is fundamental. Whenever implementing a process, a mechanism that can ensure or at least helps smooth execution of the process is a must. The more *spontaneously* the mechanism can make the process flow *automatically*, the better. As an interesting example, we can think about the hotel room access cards that are designed to be used not only for opening the door but also is to be inserted into the card slot at the door entry to turn on the electricity in the room. Without the room key card being inserted into the slot, the electricity will be off. Therefore, when the guests leave the room with their cards with them, the electricity in the room *automatically* goes off within a few seconds. Without that self-regulated mechanism (i.e. using the card to control), no matter how many signs posted in the room to remind the guests to turn off the lamps, there will always be some guests who simply ignore the request and leave the lights on when leaving the room. This example of hotel access card illustrates the importance of building a robust process that does not completely rely on the compliance of people, but focus more on the underlying mechanism of the operation. Now thinking about the analytical development, even if the team members all know that teamwork is essential and frequent communication is beneficial, it is still possible that different chromatographic methods are used by different analytical scientists who work on the same APIs but in different formulations under other projects. In an extreme case, the scientists conduct their own API stress studies, API-Excipient compatibility studies, API–API compatibility studies (for products containing multiple APIs), and API–Placebo compatibility studies, etc. From the team perspective, this is not an efficient way of utilizing the team resources. Furthermore, since the methods used in the studies are different, not only the work is unnecessarily repeated but also the knowledge gained is possessed only by the scientists who use their own methods to conduct the work. To enable the information flow automatically, to make the knowledge sharing become automatically, a mechanism of bringing the teamwork on the same platform is essential.

The good news for the analytical scientists who work in the OTC industry is that the number of molecules that are frequently used as the active pharmaceutical ingredients in OTC products is not that large. The chemical structures of some well-known APIs are listed in Table 2.2 as examples.

Table 2.2 Common active pharmaceutical ingredients found in OTC products.

Name	Molecular structure
Acetaminophen (Paracetamol)	
Aspirin	
Benzocaine	
Cetirizine	
Chlorpheniramine	
Clotrimazole	
Dextromethorphan	
Diphenhydramine	

(Continued)

Table 2.2 (Continued)

Name	Molecular structure
Famotidine	
Fexofenadine	
Fluticasone	
Guaifenesin	
Hydrocortisone	
Ibuprofen	
Levonorgestrel	

Name	Molecular structure
Loperamide	
Loratadine	
Naproxen	
Oxymetazoline	
Pseudoephedrine	
Ranitidine	

The limited number of APIs provides a unique opportunity for analytical development. The team can work together and develop a few chromatographic procedures with tremendous separation power for degradation product analysis using the QbD approaches depicted in Figure 2.3. The team then uses those procedures as platform methods for analyzing various samples generated during forced degradation studies, excipient compatibility studies, and prototype stability studies for different projects. In other words, those platform methods serve as the vehicles that carry the obtained knowledge. This approach makes learning automatically sharable among the analytical team.

Obviously, it is not possible to have one method to handle the analyses of all the degradation products of all APIs. The methods can be developed based on APIs in different therapeutic categories. The goal is that one method serves as the base method for all degradation products of all APIs that belong to one category, such as allergy, analgesics, cough and cold, etc. The methods by no means can separate all the potential degradation peaks, excipient peaks, system peaks, etc. in all existing or potential formulas. The methods may do about 80% of the job and the rest 20% imperfection of the method has to be tolerated by the analytical scientists when using those platform methods for product development. Indeed, being able to tolerate the imperfection is challenging, but it is a needed skill for the analytical scientists, who are detail-oriented and tend to pursue perfection. The methods are mainly used in preformulation and formulation studies before the formula is locked. Approaches listed below can be helpful to enhance flexibility and to help the analytical scientists tolerate that 20% imperfection:

- At the preformulation stage, use placebo as control and subtract placebo peaks instead of trying to separate the interfering placebo peaks because it is possible that in the end, the interfering excipient(s) may not find their way into the final formula;
- Employ photodiode array detectors and use different monitoring wavelengths to minimize interferences of non-API related peaks at the early stages of the formulation development, and only determine the final monitoring wavelength after the formula is locked;
- Apply less system suitability requirements to shorten the experimental run time;
- Do not focus on peaks, for example, <0.05%, unless there is a structure alert of genotoxicity of the identified degradation compound or the peak size grows fast during the prototype stability assessment;
- If the method must be optimized for a unique prototype/formula, then modify the method behind the scene by other team members while the front line lead scientist still maintains the pace of product development (teamwork);
- Stress the prototypes early on to prepare for encountering potential difficulties in the future when conducting extraction of aged samples, instead of waiting for three-month or six-month 40°C/75%RH storage;
- Use multiple methods only if by doing so can save time for data processing and interpretation.

In parallel to the development of degradation product analysis methods, some *universal methods* can be developed for assay, content uniformity, blend uniformity, and dissolution analyses. A universal method is intended to handle potency analysis of most, if not all, of the APIs the analytical development team will encounter during daily work. During the formulation development, especially at the early stages, evaluations on formulation blend uniformity, content uniformity, and quick reads on dissolution profiles are usually in high demand. It is hugely beneficial to have a method that can analyze the potencies of most of the APIs encountered during formulation development. Therefore, approaches such as applying one or a few universal assay/dissolution/content uniformity/blend uniformity methods that retain as many APIs as possible in one method are worth considering. For example, USP draft chapter <321> Drug Product Assay Tests – Organic Chemical Medicines, describes one such method that provides an HPLC assay procedure using photodiode array ultraviolet spectroscopic detection for the identification and quantitation of the following APIs in drug products.

1) Acetaminophen
2) Aspirin
3) Brompheniramine maleate
4) Dextromethorphan hydrobromide
5) Chlorpheniramine maleate
6) Doxylamine succinate
7) Diphenhydramine citrate
8) Diphenhydramine hydrochloride
9) Pseudoephedrine hydrochloride
10) Phenylephrine hydrochloride
11) Caffeine

As a side note, to facilitate the free sharing of the knowledge among the team members, besides building the mindset of teamwork, modernization of the laboratory infrastructure is also essential. Technologies such as Electronic Laboratory Notebooks and well-built user-friendly Laboratory Information Management System are necessary to serve the purpose of free knowledge sharing. More detailed discussion on the modernization of analytical laboratory digital infrastructure will be beyond the scope of this book, but interested readers are encouraged to study further.

As a disclaimer, the above discussions and approaches in conducting analytical development work only serve for information purpose. Other approaches unquestionably exist, and the readers are encouraged to develop different methodologies.

2.6 Methods for Scale-Up and Manufacturing QC Laboratories

After preformulation and formulation development, the product development moves forward with the most promising laboratory scale formula. Scale-up and/or technology transfer (TT) team in a product development setup can exist either as part of an integrated formulation-process group or as a separate entity apart from the formulation development. The technology transfer group determines (1) the scalability of formulation prototypes developed at the laboratory scales, (2) the integrity/continuity of the performance characteristics designed for the formula when large batch sizes are attempted, and (3) the robustness of the manufacturing process. The technology transfer scientists transfer the formulation, process, and knowledge to the manufacturing site(s).

According to the FDA's Scale-Up and Post-Approval Changes guidelines, the entire scale-up process must be validated whenever the process is scaled-up by a factor of at least 10 [26–29]. Indeed, the path from laboratory-scale batches to production-scale batches typically involves factors of 10 or more of the initial amounts. During the technology transfer, the analytical development scientists work closely with the formulation scientists, engineers, and analysts in the quality control laboratory of the manufacturing site. Content uniformity, dissolution profiles, and stability of the manufactured products are extensively tested at the analytical development laboratory or the manufacturing site QC laboratory. In the meantime, the analytical methods are finalized and validated/transferred to the manufacturing site QC laboratory. In case of process changes due to the scale-up of the production or due to some discrepancy found in the product performance between laboratory prototypes and manufacturing scale batches, the analytical development scientists who have been working with the product development will also conduct extensive analytical work to support the manufacturing process and/or formulation investigations.

At the end of the laboratory-scale formulation development, the composition of the formula will be finalized. From the QbD point of view, the corresponding analytical methods then need to move from the platform methods (Knowledge Space methods) to methods that are more specific to the formula (Design Space methods, Figure 2.8). Again, note that the knowledge space is built for the entire analytical development team. The knowledge space is not built for one person or one project. The number of chromatographic peaks monitored by the Design Space methods may be less than the peaks monitored by the platform methods based on the knowledge gained. The analytical methods can thus be shortened. Supposedly there can be a third stage of method development, that is, building Control Space methods. It is the finalization of the methods so they can be validated and transferred to QC laboratories for routine use. Interestingly, however, although the terms Design Space and Control Space, and the intention of creating such spaces make perfect sense for manufacturing process development, it is not necessary to separate the analytical method finalization/optimization into separate Design Space and Control Space. With nowadays column technologies, such as using sub 2 μm ultrapure silica particles as column packing, the methods developed for the Design Space can possess sufficient separation power together with short run times. It actually makes more sense to

Build knowledge space – get AD team ready

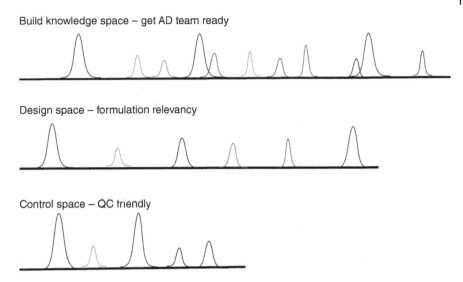

Design space – formulation relevancy

Control space – QC friendly

Figure 2.8 Knowledge space, design space, and control space of method development.

regard the Design Space as the Robustness Space of the final methods, rather than a separate phase of method development.

When finalizing a new method for a new product, the analytical development scientists should keep in mind that the purpose of a method for R&D and the purpose of a method for a QC laboratory are very different. The former one is developed when racing against time to generate (a large amount of) data to provide insightful information for the formulation development, the latter one is to be used routinely for at least a few years until the life cycle of the product ends. Since the final users of the methods are QC laboratories, the procedures developed by the analytical development team need to be polished so the methods can be transferable and are QC friendly. There are many *want* and *don't want* in an analytical procedure that QC laboratories will have to use day-in and day-out. For example, even a simple practice of pH adjustment by adding drops of acid or base into the buffer solution is not that easy and straightforward within the context of a QC laboratory setting. Some QC laboratories do not allow direct immersion of the pH probe to the bulk of the buffer solution when adjusting the pH to avoid cross-contamination. The QC analyst has to pour a small portion of the buffer solution to a separate beaker to measure the pH and to repeat the steps of "adding acid or base – stirring the bulk buffer solution – pouring a small portion of the buffer solution to a separate beaker – measuring the pH to confirm." Furthermore, it is hard to keep consistency in laboratory practices between different laboratories, if the details of procedures are not clearly spelled-out. Inconsistencies in some key parameters could then create inconsistency in method performance. Therefore, it can be even more time consuming for the analytical scientists in the analytical development to finalize/fine-tune/polish the methods than to develop the methods. Certain character traits preferred in an HPLC method that is intended for QC use include the following:

- Short run time
- Column temperature controlled
- Robust and Rugged
- Reproducible on any brand of HPLC systems
- Long column life time
- Baseline separation of all named impurities
- Easy on peak integration/quantitation
- Mobile phase compatible with LC-MS
- At least two brands of columns that are true equivalent (a stretch preference)

The items that are not preferred in the methods intended for QC routine use include the following:

- Methods that require QC analysts to adjust experimental conditions to meet system suitability requirements
- Filtration of mobile phases after mixing of organic solvents
- Instrument specific
- Vendor specific reagents, solvents, etc.
- Using THF in the diluent or mobile phases

There is no clear cut when method development should be moved from building the Knowledge Space to building the Design/Control Space. A good timing for such a transition can potentially be found when the product development moves to the scale-up stage and ready to be transferred to manufacturing sites. Sometimes, three-month or six-month informal stability studies, containing storage of samples under accelerated conditions (such as 40°C/75%RH), are being conducted at this stage. The analytical scientists can take the advantage of the time (three to six months) and the available stability samples to switch from using the platform methods to developing formulation specific methods, and optimizing/fine-tuning the methods so that the methods can be suitable for routine use in QC laboratories. To some extent, many QbD approaches reported in literature seem to be applicable for method development at this stage.

Some companies have two analytical groups, one is the method development group and the other one is the stability-testing group. The method development group is responsible for developing the method, and the stability-testing group is responsible for validating and transferring the method to QC laboratories or to conduct in-house stability testing. It is not necessarily better in one way or the other, but although having two separate groups can reduce the workload of individual team members, it can also create silos and miscommunications between the two groups. The method development team may not pay that much attention to the method details since they are not the ones who will routinely use the method they developed. The stability group, on the other hand, may not have time (or may not be given sufficient time) to fine-tune or polish the method, or do not have sufficient information on the development history of the method that is handed over to them. Gap in terms of the understanding of the method nuisances can form between those two groups. It is in the author's opinion that the method developer should be responsible for the entire life cycle of the method, from the development to the routine use of the method in QC laboratories.

After having finalized the methods, the analytical scientists start working on method validation and transfer. No matter how you look at it, it seems the biggest challenge that an analytical scientist faces in the pharmaceutical industry, is the analytical method validation and transfer. There are conferences specifically focus on that topic and it is a critical component of the registration documents that get submitted to health authorities. Method validation and method transfer somehow give people the notion that those activities mean that there is no room for errors, and the most nerve-racking thing is, statistics are now involved in the result analysis. However, an analytical scientist who does not want to do things in a black box has to study some statistics applied in analytical chemistry. Some basic knowledge of statistics will be presented in Chapter 6.

The work is, however, not complete after the method validation and transfer. Life cycle management of an analytical method has recently become a popular topic in the analytical society. The changes in column surface chemistry, changes in suppliers of the API, changes in synthetic routes of the API, etc. can potentially render the validated method not working anymore. Sometimes when columns get aged, the chromatographic peak retention pattern can have some slight changes, which may impact a critical pair of degradation product peaks. Some companies request two HPLC columns to be validated at the same time, one works as the primary and the other one as the backup column. Ironically, the author has experienced that the backup column went out of the market earlier than the primary column! Therefore, an analytical scientist who develops the method has to take the responsibility to provide continuous support to troubleshoot the method even after the method has been transferred out of his/her hands.

2.7 Process Analytical Technology

Besides supporting formulation development, analytical development can potentially engage in the development of the manufacturing process. One topic worth mentioning in the framework of the manufacturing process, especially within the scope of Integrated Continuous Manufacturing (ICM), is the concept, application, and challenges associated with the so-called process analytical technology (PAT). There are abundant literature and websites for interested readers to read and learn in more detail concerning the PAT. This book will not conduct deep discussions regarding the technologies and will only provide a brief high-level overview.

In its draft guideline "Quality Considerations for Continuous Manufacturing" published in February 2019, the FDA defines a continuous manufacturing process as a process consisting of at least two connected Unit Operations to which material is continuously fed and from which material (product) is continuously removed. The FDA recognizes continuous production as the most important tool in the modernization of the pharmaceutical industry and considers the concept closely related to the agency's QbD initiative. A critical component that spans the entire ICM is the PAT. As early as in 2004, the FDA launched the PAT concept to stimulate the pharmaceutical industry to change from off-line to real-time quality testing. In 2014, the FDA published "Guidance for Industry PAT – A Framework for Innovative Pharmaceutical Development, Manufacturing, and Quality Assurance," in which the concept of Analytical in PAT is broad and includes aspects of chemical, physical, microbiological, mathematical, and risk analysis. To ensure the proper use of PAT in ICM, the analytical development team needs to conduct a large amount of prework to establish the mathematical models. A thorough understanding of the manufacturing process is a must. Multivariate methodologies and statistical analyses need to be conducted to extract critical process knowledge for real-time control and quality assurance. Good correlations between the PAT data and the conventional chromatographic data must be achieved.

As of 2020, when this book is written, spectrometry (Figure 2.9) is still the main analytical tool for in-process evaluation.

Near-infrared (NIR) spectroscopy and Raman spectroscopy are mostly used because of their speed of analysis and less labor-intensive sample preparations. To an analytical scientist, PAT means vibrational spectroscopy + chemometrics, or in other words, the PAT means NIR or Raman + statistics. However, simply speaking, chemometrics is not another name of statistics. Chemometrics is concerned with the application of mathematical and statistical techniques, with the help of computer science, to extract chemical and physical information from complex data. A statistician may not know about chemistry or instrumentation, while a chemometrician brings knowledge of the chemical and sometimes the instrumental influences, which affect the data. The aim is often to display the data in ways that allow chemical interpretation of the system.

NIR and Raman are widely used in manufacturing and quality control laboratories for raw material identification, finished product identification, polymorphism

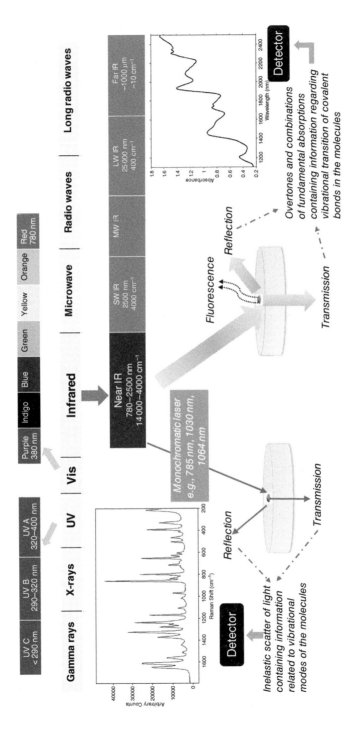

Figure 2.9 Near-infrared and Raman spectroscopy at a glance.

characterization, API and excipient quantification, etc. The popularity of NIR and Raman techniques comes from the following features:

- Deeper penetration depth (1–3 mm compared to 0.5–2 mm in ATR-IR, attenuated total reflectance infrared spectroscopy), larger sample volume interrogated (10 mm^3 for Raman, 75 mm^3 for NIR). Both of which enable a large sample amount to be analyzed with one measurement and thus enhances the data reliability, especially for identification purpose.
- Convenient, potable, or handheld, nondestructive, and can be used for in situ analyses.
- No or little sample preparation or solvent is required.
- Spatially offset Raman spectroscopy can direct analyze through opaque packaging.

Complexity associated with the application of Raman includes the interference from fluorescence and lack of sensitivity. The use of laser wavelength at 785 nm in common Raman devices is to minimize the fluorescence interferences and at the same time to maintain needed sensitivity. As for NIR, the biggest hurdle of its application for quantitative analysis is the complicated interpretation of spectra data that may need highly trained staff to build chemometric models. Furthermore, probe contamination is a critical challenge for in-line NIR or Raman application in manufacturing process monitoring. Unclear sample size (e.g. the sampling volume of powder measured during the collection of a spectrum) is another known challenge. The scientists have to build robust and accurate calibration models based on the products to be made, in similar circumstances and with equipment chain (a chicken-or-egg situation). Spectral databases or models may not be directly transferable to different brand of devices possessed at different manufacturing sites.

PAT in the manufacturing process, therefore, is still under development. There might be still a long way to go. Moreover, the technical skill sets needed for PAT are quite different from what is possessed by the conventional separation scientists. An analytical development team may need to expand its talent pool and embrace new technologies.

2.8 Quality Assurance, Compliance, and Analytical Development

Quality assurance is a broad concept that covers all aspects that collectively or individually impact the quality of the product. It is an obligation for the manufacturers to meet the needs of end user in terms of safety, quality, efficacy, strength, reliability, and durability. According to ICH Q10, the pharmaceutical quality system "assures that the desired product quality is routinely met, suitable process performance is achieved, the set of controls are appropriate, improvement opportunities are identified and evaluated, and the body of knowledge is continually expanded." Based on the FDA Guidance for Industry "Quality Systems Approach to Pharmaceutical Current Good Manufacturing Practice Regulations," QA primarily involves in (1) reviewing and approval of all procedures related to production and maintenance, (2) reviewing of associated records, and (3) auditing and performing/evaluating trend analyses. The cGMP regulations specifically assign the quality unit in a company the authority to create, monitor, and implement a quality system. Such activities do not substitute for, or preclude, the daily responsibility of manufacturing personnel to build quality into the product. The QA unit should not take on the responsibilities of other units of a manufacturer's organization, such as the responsibilities handled by manufacturing personnel, engineers, and development scientists. Other cGMP assigned responsibilities of the QA unit are consistent with modern quality system approaches, such as

- Ensuring that controls are implemented and completed satisfactorily during manufacturing operations.
- Ensuring that developed procedures and specifications are appropriate and followed, including those used by the company under contract to the manufacturer.
- Approving or rejecting incoming materials, in-process materials, and drug products.
- Reviewing production records and investigating any unexplained discrepancies.

Depending on the organizational structures, some R&D analytical development departments may have dual responsibilities, one for research, experimental, exploratory work, and one for GMP work that is focused on compliance. Examples of the work related to GMP activities include releasing raw materials and clinical supplies for clinical trials, certifying reference standards for method validation, conducting stability testing for product registration, carrying out experiments to fulfill health authority's requests. To ensure compliance when conducting GMP work, the operation of the laboratory must be examined from the regulation compliance point of view although it may require lots of documentation and infrastructure support, that apparently may slow down the data/result turn around. A healthy balance between research and GMP mindsets, and adequate management support from laboratory space (to physically separate research and GMP activities as much as possible), validated electronic system to ensure data integrity, and financial aspects are imperative for the analytical development laboratory to maintain the productivity.

References

1 Pharmaceutical Development Q8(R2) (August 2009). The International Council for Harmonisation of Technical Requirements for Pharmaceuticals for Human Use (ICH).

2 The U.S. Food and Drug Administration (FDA) (March 2007). Guidance for Industry and Review Staff Target Product Profile — A Strategic Development Process Tool.

3 Quality Risk Management Q9 (November 2005). The International Council for Harmonisation of Technical Requirements for Pharmaceuticals for Human Use (ICH).

4 https://www.fda.gov.

5 The U.S. Food and Drug Administration (FDA) (April 2004). Guidance for Industry Changes to an Approved NDA or ANDA, Revision 1.

6 Covey, S.R. (2013). *The 7 Habits of Highly Effective People*. Simon & Schuster, 25th Anniversary Edition.

7 European Medicines Agency (WMA) (November 2017). Guideline on the Requirements to the Chemical and Pharmaceutical Quality Documentation Concerning Investigational Medicinal Products in Clinical Trials, Revision 1.

8 Bansal, A.K. and Koradia, V. (2005). The role of reverse engineering in the development of generic formulations. *Pharmaceutical Technology* 29 (8): 50–55.

9 Impurities in New Drug Products Q3B(R2) (June 2006). The International Council for Harmonisation of Technical Requirements for Pharmaceuticals for Human Use (ICH).

10 The U.S. Food and Drug Administration (FDA) (July 2006). Guidance for Industry Q3B(R2) Impurities in New Drug Products.

11 The U.S. Food and Drug Administration (FDA) (November 2010). Guidance for Industry ANDAs: Impurities in Drug Products.

12 Impurities in New Drug Subtances Q3A(R2) (October 2006). The International Council for Harmonisation of Technical Requirements for Pharmaceuticals for Human Use (ICH).

13 The U.S. Food and Drug Administration (FDA) (June 2008). Guidance for Industry Q3A Impurities in New Drug Substances.

14 https://www.usp.org.

15 United States Pharmacopeia (USP) General Chapter <701> Disintegration. The United States Pharmacopeial Convention. Official 1 May 2020.

16 Al-Gousous, J. and Langguth, P. (2015). Oral solid dosage form disintegration testing — the forgotten test. *Journal of Pharmaceutical Sciences* 104: 2664–2675.

17 Sengupta, P., Chatterjee, B., and KumarTekadea, R. (2018). Current regulatory requirements and practical approaches for stability analysis of pharmaceutical products: a comprehensive review. *International Journal of Pharmaceutics* 543 (1–2): 328–344.

18 Chaurasia, G. (2016). A review on pharmaceutical preformulation studies in formulation and development of new drug molecules. *International Journal of Pharmaceutical Science and Research* 7 (6): 2313–2320.

19 Lee, R.W. and Mitchnick, M. Early-stage formulation considerations. *Current Protocols in Chemical Biology* 9 (4): 306–314.

20 Honmane, S.M., Dange, Y.D., Osmani, R.A.M., and Jadge, D.R. (2017). General considerations of design and development of dosage forms: pre-formulation review. *Asian Journal of Pharmaceutics* 11 (3): 479–488.

21 Li, J., Lee, J., Varanasi, M., and Xiao, K.P. (2012). Improved analytical recovery by taking into account sample matrix and chromatographic instrumentation. *American Pharmaceutical Review* 15 (7): 32–36.

22 Gray, V.A. (2018). Power of the dissolution test in distinguishing a change in dosage form critical quality attributes. *AAPS PharmSciTech* 19 (8): 3328–3332.

23 Song, X., Cui, Y., and Xie, M. (2011). Gelatin capsule shell cross-linking. Tier II dissolution method development in the presence of sodium lauryl sulfate. *Pharmaceutical Technology* 35 (5): 62–68.

24 The U.S. Food and Drug Administration (FDA) (August 1997). Guidance for Industry Dissolution Testing of Immediate Release Solid Oral Dosage Forms.

25 European Medicines Agency (EMA) (August 2010). Guideline on the investigation of bioequivalence (Revision 1).

26 The U.S. Food and Drug Administration (FDA) (November 1995). Guidance for Industry SUPAC-IR: Immediate-Release Solid Oral Dosage Forms: Scale-Up and Post-Approval Changes: Chemistry, Manufacturing and Controls, In Vitro Dissolution Testing, and In Vivo Bioequivalence Documentation.

27 The U.S. Food and Drug Administration (FDA) (October 1997). Guidance for Industry SUPAC-MR: Modified Release Solid Oral Dosage Forms Scale-Up and Postapproval Changes: Chemistry, Manufacturing, and Controls; In Vitro Dissolution Testing and In Vivo Bioequivalence Documentation.

28 The U.S. Food and Drug Administration (FDA) (May 1997). Guidance for Industry SUPAC-SS: Nonsterile Semisolid Dosage Forms; Scale-Up and Post-Approval Changes: Chemistry, Manufacturing and Controls; In Vitro Release Testing and In Vivo Bioequivalence Documentation.

29 The U.S. Food and Drug Administration (FDA) (December 2014). Guidance for Industry SUPAC: Manufacturing Equipment Addendum.

3

Effective, Efficient, and Innovative Analytical Development

In a fast-paced pharmaceutical product development environment, being an analytical scientist that is chosen to be part of a research and development project team can be very exciting. However, to be a part of a project team that consists of members from various functional areas can also be challenging and stressful. For example, the lead analytical scientist has to manage various complexities in addition to analyzing a large number of formulation samples. The analytical scientist has to do the work effectively and efficiently and do it by collaborating with other team members. The scientist needs to constantly think how to manage the heavy workload, how to prioritize the work, how to juggle among different projects that he/she is responsible for, how to work with various colleagues and team members who have vastly different personalities and who have their own priorities, how to negotiate for more support from the management, etc. From the technical aspect, the lead analytical scientist has to think how to switch between conducting research work and performing work for GMP purposes, how to balance the necessity to carry out, for example dissolution tests versus accepting quick results based on disintegration evaluation, when to perform single-point dissolution versus obtaining dissolution profiles, when to get a good enough assay value versus getting a value with perfect accuracy, when to monitor every single impurity peak observed on chromatograms versus focusing on the critical ones, etc. However, the lead analytical scientist does not make the decisions just by playing solo. In a team environment with various functions, assertive communications, seeking to understand before being understood, active listening, etc. are very important soft skills that a lead analytical scientist needs to have. And, of course, characters such as confidence, believing in him/herself, expressing opinions with a confident attitude are very important, which can only come with a clear mind and a solid understanding of the work process.

An analytical scientist to some extent by default is very much detail-oriented and data-driven. We can sometimes be a little stubborn, or even rigid to stick to

Analytical Scientists in Pharmaceutical Product Development: Task Management and Practical Knowledge, First Edition. Kangping Xiao.

the principles that we believe. We believe that "The data is data," which means sometimes we have to be the messenger who bears the bad news. However, analytical scientists need to make efforts to dissolve into a collaborative team environment, although not by bending the quality and integrity of the analytical work. There are many tools people can find to help manage the complexity. This chapter introduces some task management tools such as fishbone diagrams and time-bars. Those tools are only for examples to provoke thinking, while not to be imposed on the readers. The chapter also is going to talk about some project management approaches, such as Gantt Chart, which links all product development activities or steps in a Waterfall fashion; and Agile project management, which contains methods like daily scrum and sprint review. A few spreadsheet templates for resource estimation are presented as well. Many other approaches and self-improvement tools can be found through an internet search or from reading books. Some companies may have internal training or can send their employees to attend relevant training courses. This chapter only presents some approaches that can be useful for managing analytical development work. The readers can self-explore and learn further if the topics are found interesting.

Some organization setup may have a Project Lead role that is played by senior-level analytical scientists or someone who has more project management skills. The project leads will attend project meetings and supervise bench scientists on the laboratory work. It is the author's opinion, however, that an analytical project lead should spend sufficient time and make a sincere effort to carry out some laboratory bench work. Keeping a close connection between the project management and the physical laboratory work is essential for the project lead to create and maintain a true partnership between the lead and junior level scientists. Moreover, when doing research and development work, the ability to observe the experimental phenomenon, and the ability to accurately estimate the resources needed for task completion is critical for the success of analytical support. Seeing the experimental happenings through their own eyes, and working on tasks by their own hands, can significantly help the project leads to develop user-friendly methods, troubleshoot issues, and to arrange resources reasonably. The hands-on experience also helps the project leads to conduct realistic estimation on long-term resource requirement and time needed to complete a project.

3.1 Task Management by Fishbone Diagrams and Time-Bars

As the saying goes, the devil is in the details. And the analytical work is always concerned about details. When planning the analytical work, it can be very easy to miss some tasks that may turn out to impacting the overall project completion.

Furthermore, since we are all asked to do more with fewer resources, people tend to do things in parallel, without realizing that a true doing things in parallel does not exist without careful planning and discipline in executions. The phrase Work in Process (WIP) is used to refer to the ongoing work, potentially consuming some resources, and yet not been completed. With too many WIP in place, a disastrous domino effect is a potential risk that should be recognized.

The analytical scientists should know what is needed to get things done. All the details of the plan and the status of plan execution should be clearly laid out in front of us and can be checked on a daily basis. Different people have different preferences in terms of how to plan and monitor the executions of the plan. Many people make Task List or To-Do List and add items to the list on the go or cross out items from the list when the task is complete. In this section, some other potential tools are presented for readers to consider.

One of the core functions a lead analytical scientist performs daily is to manage analytical work/activities associated with assigned projects. The analytical development interacts with many departments (Figure 3.1). For example, the analytical development works closely with the formulation development (FD) and scale-up and technology transfer colleagues to create new products; works with the scale-up and technology transfer team to develop a robust manufacturing process; and works with the clinical operations to facilitate clinical trials. The analytical

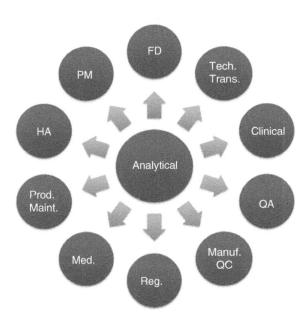

Figure 3.1 Analytical development interacts with various functions.

development interacts extensively with the quality assurance department. The methods developed by analytical development are mostly used in the manufacturing quality control (QC) laboratories. The methods are developed with the consideration of the regulatory requirement in mind, and the analytical development contributes a large part to the compilation of the registration documents for regulatory filing. The medical affairs colleagues may ask the analytical development team to provide dissolution profile comparisons between the new product and some existing products for bioequivalence estimation. During the life cycle management of the product, i.e. product maintenance, analytical development may from time to time get involved in the identification of new unknown peaks observed or to help out-of-specification investigations. When the company receives deficiency letters from the health authorities (HA) regarding the approval of a new product or for rolling-out to other markets of some existing products, the analytical development often gets involved and has to provide answers within a tight timeline. And, project managers (PM) will constantly come to the lead analytical scientist for status updates and to seek alignment.

To manage all the relationships, tasks, and timelines, the analytical scientist thus has to have a very clear understanding of what his/her roles and responsibilities are, and how to prioritize the work.

Many tools are available through the Internet, company training, or books to help manage all the tasks. In this book, Ishikawa diagrams (commonly known as fishbone diagrams) are used to help visualize the tasks and their relationship. The way to use those fishbone diagrams here is somewhat different from its traditional application. People use the fishbone diagrams to list the *causes and effects* when conducting troubleshooting. But using fishbone diagrams to manage tasks can be very flexible and effective. The fishbones drawn in the following diagrams are by no means all-inclusive. The readers are encouraged to use these diagrams to manage tasks and projects in more creative ways. The end goal of creating a final fishbone diagram is that each fishbone on the diagram represents an executable task.

Figure 3.2 is a "fish" that is called "analytical development." That is a big fish, a "shark." It is the very first level of the diagram that a lead analytical scientist draws. It represents the understanding of the overall roles and responsibilities the analytical development possesses in a product development project. The analytical scientist knows that Method Development-Validation-Transfer activities are the core analytical tasks. The analytical work carried out for supporting formulation development, scale-up and technology transfer is the reason why the analytical scientist is on the project team. Providing high-quality analytical documents to help regulatory filing is the final closure of the analytical work during product development. The product life cycle management requires the analytical scientist who participated the original development to help troubleshoot the method performance issues encountered during the routine use of the method, or to carry out impurity identification should some unknown peaks are observed during the

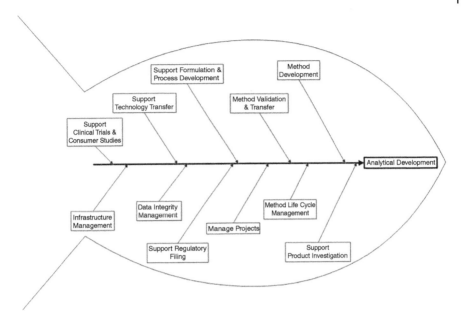

Figure 3.2 Level I fishbone diagram depicting the main elements of analytical development.

commercialization of the product, and to provide answers from various health authorities regarding the analytical methodology and stability data. To ensure high-quality analytical work, particularly the work for supporting GMP activities (such as clinical trials), the lead analytical scientist makes sure the data integrity, clarity and accuracy of the notebook documentation, calibration status of the instruments, the authenticity of the reference standards, etc. Therefore, the lead scientist needs to help the team to build or maintain a robust and modern infrastructure (such as using electronic systems rather than relying on paper systems) and to maintain compliance with regulations. The lead scientist knows that without collaboration with colleagues in the analytical development department and with colleagues in all the other functions, and sometimes without working with the partners or service providers from the contract organizations, he/she can never achieve all the goals that are necessary to fulfill the duties. To be able to manage all those tasks in an organized way, the lead analytical scientist also needs to be familiar with project management.

The importance of Figure 3.2 is that it paints a big picture in front of the lead analytical scientist by describing things at a philosophical level. Each fishbone in Figure 3.2 contains a large amount of information, and each one of them by itself is a big topic. We can label this giant fish as the "Level I fishbone diagram." For example, Figure 3.3 presents a fish called "Method Development," which is an expansion of one of the fishbones in Figure 3.2. Note here the method (by default)

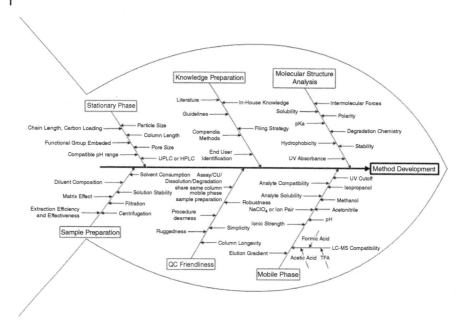

Figure 3.3 Level II fishbone diagram depicting the main elements of method development.

refers to a chromatographic method. Several aspects of a method development such as "Molecular Structure Analysis," "QC Friendliness," "Sample Preparation," etc. are the big fishbones, and "Polarity," "Simplicity," "Matrix Effect," etc. are the smaller fishbones on the respective big fishbones. The fishbone diagram in Figure 3.3 is a busy diagram since it starts to contain some details. We can label the chart at this level of detail as a "Level II fishbone diagram." Different scientists may regard different parameters as big or small fishbones. Different projects or different stages of a project can have different priorities, and therefore, the same parameters can be big or small fishbones on different fishbone diagrams. The good thing about building a warehouse of this kind of fishbone diagrams is that they can be modified (electronically on a computer) as much as needed and can be reused for other projects as many times as needed.

Although Figure 3.3 is already busy with many smaller fishbones, this "method development fish" is somewhat still at a philosophical level. The situation is similar in Figure 3.4, which is a "Method Validation Fish." The fishbones such as "Acid" "Base" on the big fishbone of "Specificity" still serve mainly as reminders for the analytical scientists to carry out those tasks, rather than telling the analytical scientist exactly what to do and when to do. A further detailed level is needed to turn the fishbones into ***executable action items***.

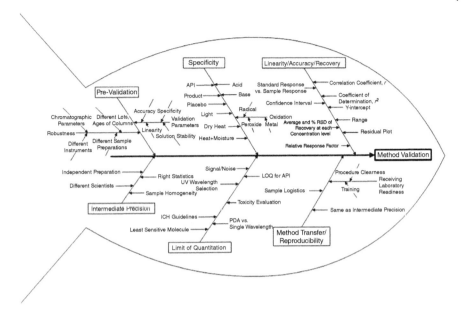

Figure 3.4 Level II fishbone diagram depicting the main elements of method validation.

Figure 3.5 shows a fish called "Knowledge (Information) Preparation." It is a fishbone diagram that is created based on one of the big fishbones on the "Method Development Fish" (Figure 3.3). In Figure 3.5, each small fishbone represents an action item, with a timeline associated with it. We label the chart at this level of detail as the "Level III fishbone diagram."

The above fishbone diagrams are for illustration purposes and are only shown for example. The "fish," the "fishbones," and the action items are meant to be changed according to the real needs in actual situations. As we can see below in Figure 3.6 through Figure 3.9, when it comes to creating fishbone diagrams based on the big fishbones of Infrastructure Management and Data Integrity Management, the levels can quickly go to Level IV or even Level V, before it reaches to the executable action items. As shown in Figure 3.6, the Infrastructure and Data Integrity chart is a level II fishbone diagram which contains many fundamental requirements by law and/or by health authority guidelines. The instruments must be kept in suitable status by being periodically calibrated and maintained. The SOPs are periodically reviewed and updated when necessary. The user privileges of the electronic systems are periodically reviewed and controlled. The service contracts and quality agreements are in place with necessary audits being conducted. The laboratory safety, housekeeping, and waste management are properly managed. The data integrity is ensured by following the

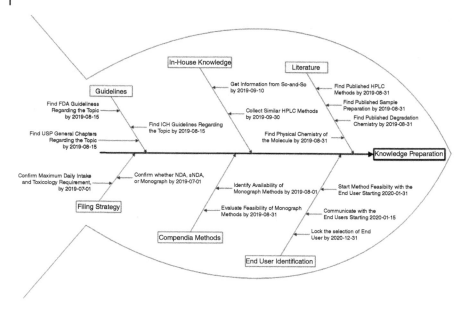

Figure 3.5 Level III fishbone diagram depicting the executable action items for knowledge (information) preparation.

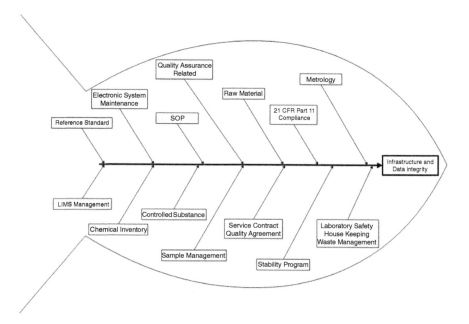

Figure 3.6 Level II fishbone diagram depicting the main elements of infrastructure and data integrity management.

requirements outlined in the 21 CFR Part 11. The controlled substance is securely stored and the usage is closely tracked. The stability program, sample management, and chemical inventory program are running smoothly behind the scene. The Laboratory Information Management System works properly to help track the sample or material flow from arriving at the building to going into the finished product. One valuable insight that Figure 3.6 illustrates is the necessity of having a separate infrastructure team to handle all those tasks. To further clarify this important point, Figure 3.7 through Figure 3.9 show the subsequent levels of the fishbone diagrams for managing a stability program, which is only one of the big fishbones on the "infrastructure fish."

Figure 3.7 illustrates a new fish that is created based on the fishbone of the stability program on the "Infrastructure and Data Integrity Fish" (Figure 3.6). On this "Stability Program Fish," many fishbones are added to the diagram that is needed to manage a stability program. Again, those items are for example and can be certainly replaced by the ones the stability coordinator sees fit. This fishbone diagram lists many items that mainly serve as reminders to ensure no tasks are inadvertently forgotten or "falling through the crack." Note Figure 3.7 is already at the level III of the fishbone diagrams. However, each fishbone still is a body of information that can be further dissected.

Figure 3.8 uses the fishbone of "Documentation Completeness" in Figure 3.7 to keep going further down to the details. The example provided here shows that the

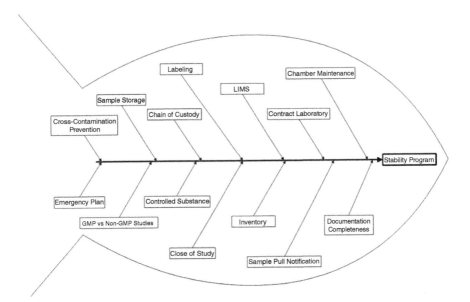

Figure 3.7 Level III fishbone diagram depicting the main elements in stability management.

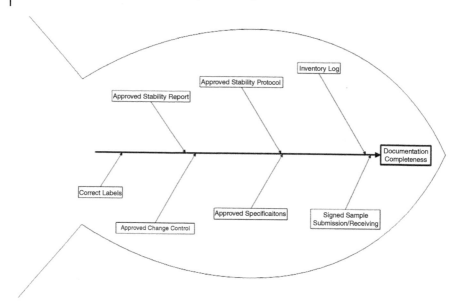

Figure 3.8 Level IV fishbone diagram depicting the main elements that are needed to ensure the completeness of documentation when managing one stability program.

stability coordinator is reminding him/herself to make sure to collect all the necessary paperwork so no document is overlooked. This fish is more like a self-reminder template for the coordinator to use whenever a stability study is going to happen.

Once the template fishbone diagram is created, the stability coordinator can then use the template to create a fishbone diagram that is specific to a certain project, Project XYZ. At this level, the executable action items start to appear on the diagram, as shown in Figure 3.9.

The above fishbone diagrams illustrate that in order to manage the "documentation completeness" task of a stability program, the stability coordinator has to spend a large amount of focused time and energy on the tasks and has to have lots of follow-up and follow-through with the team to ensure the completion of the tasks. However, although those tasks are incredibly important in a regulated industry since no company can afford to have confusing stability management or to create many quality events due to missing needed paper work, the lead analytical scientists won't have time to manage those logistic tasks since they have other equally important tasks to manage, that is, to help the team to launch quality products to the market.

Let us pause and see the difference between a "Stability Fish" created by the lead analytical scientist (Figure 3.10) and the "Stability Program Fish" that is created by the stability coordinator in Figure 3.7. The tasks important to the lead scientist are very much technically focused, while the items critical to the stability

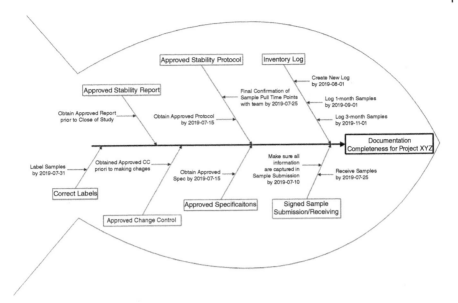

Figure 3.9 Level V fishbone diagram depicting the executable action items for completing the documentation for managing one stability program.

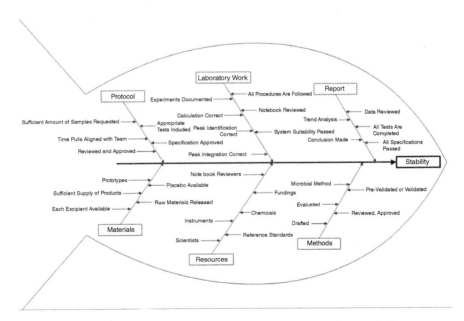

Figure 3.10 A fishbone diagram depicting the critical elements for completing a stability study from a lead analytical scientist's point of view.

coordinator are logistic and compliance-related. The difference between Figures 3.7 and 3.10 illustrates the importance of having an infrastructure team to provide logistic support to the front line analytical scientists to manage the projects. The lead analytical scientists need solid supports from the infrastructure team and need to feel assured that no gap exists in data integrity and GMP compliance. Only with strong support from the teamwork, the lead scientists can focus on science and project management.

Before going to the discussion of project management, let us take a closer look at the tasks that a lead analytical scientist needs to manage. Figure 3.11 is another diagram that helps visualize the workflow, tasks involved, and working relationship among team members. Besides developing methods and testing prototypes, which can be done to some extent independently, the other tasks carried out during formulation development are related to team interactions. The lead analytical scientist needs to work with formulation scientists to codevelop stable products; the lead analytical scientist needs to work with the project manager to estimate the full time equivalent (FTE) and costs associated with the project. The lead scientist needs to work with clinical operations to ensure the integrity of the clinical trials. Although not included in the diagram, mainly for the reason not to make the diagram overcrowded, the lead analytical scientist works with the analytical infrastructure team to make sure the tasks such as raw material release is done in a timely fashion and the reference standards are certified for GMP use, etc. Once the formula is locked, the workflow moves on to the scale-up and technology

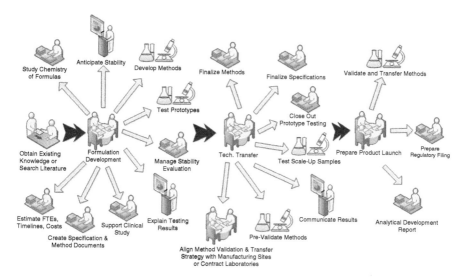

Figure 3.11 Tasks and interactions a lead analytical scientist has during a product development.

transfer group that manages scaling up of the formulation and the manufacturing process optimization. The analytical development at this point reaches to the stage of finalizing the methods and specifications. Depending on the company practices, method validation and transfer can occur at this stage as well. The interactions the analytical development will have at this stage can extend beyond the research and development (R&D) team to manufacturing sites or contract research organizations (CROs)/contract manufacturing organizations. With a successful scale-up and transfer of the manufacturing process to the designated manufacturing site(s), the lead analytical scientist will wrap up the development work and help prepare high-quality registration documents for regulatory filing.

Now, being knowledgeable of the workflow and the tasks needed to complete new product development, one critical element starts to become evident in the picture, that is, the timelines. Figure 3.12 shows a fishbone diagram that has a time-bar inside. That is an example of using a fishbone diagram to manage the analytical deliverables along with the development progress of project XYZ. The timeline in the diagram is for illustration purposes, and the tasks placed in the diagram do not reflect any real cases. So for project XYZ, the lead analytical scientist wants to make sure that the raw materials are officially released by week 36 of year 1 (the total number of weeks in a year is 52). The early prototypes, i.e. preliminary formulas, have to be analyzed by week 36 of year 1 as well. The project should have received a positive business case before the kick-off but should at least receive a go or no-go verdict in week 43, ideally with not only the market research but also maybe some consumer evaluations on the early prototypes

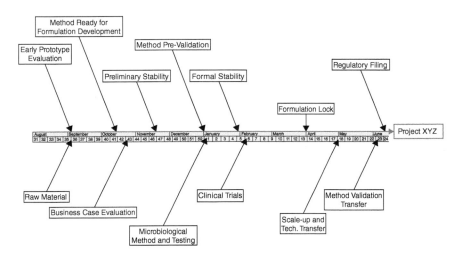

Figure 3.12 A fishbone diagram depicting the elements to be managed for completing project XYZ along with the timelines associated with each element.

regarding their appearances, tastes, flavors, and other perceptions. In the meantime, some preliminary stability studies are underway from roughly week 38 to week 50. Once the project receives a green light of moving forward, the formulation work will speed up with a variety of prototypes rushing to the analytical laboratory. To ensure the method integrity and get ready for the formal stability testing and supporting of clinical trials, the lead analytical scientist will perform some preliminary method validation work in a semi-formal manner, by week 1 of year 2. The microbial testing method is to be validated around week 1 to week 4 of year 2 so the microbiologic method can be ready for formal stability studies. The formulation is then locked by week-14 of year-2, and the project moves to scale up batch manufacturing and technology transfer to the manufacturing site for registration and commercial production. During this time, the analytical development works with the manufacturing site QC laboratories and carries out the method validation and transfer work, by week 23 of year 2. At the end of the product development, the analytical scientist helps to prepare the registration documents for regulatory filing, with a deadline in week-24 of year 2.

Again, the time-bar in Figure 3.12 is only for illustration purposes. In reality, product development teams never carry out tasks in a sequential order. Project management, time management, resource management, team collaboration, frequent communication, etc. are thus essential to the success of teamwork.

In the beginning of this section it is mentioned that one of the core functions a lead analytical scientist performs daily is to manage projects. You may have noticed that the word "projects" is a plural form. Indeed, managing multiple projects in parallel is very common in the new product development environment. To effectively manage the tasks and arrange the resources efficiently, a "Multi-Project Time-Bar" which has the shape shown in Figure 3.13 can be helpful. When

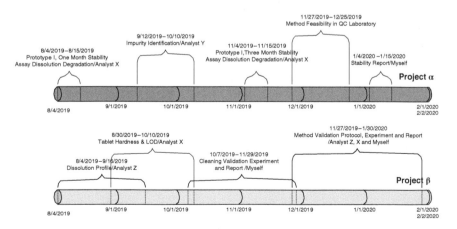

Figure 3.13 Multi-project Time-bar that can be used for resource arrangement and task execution planning.

managing multiple projects, it is not rare for analytical scientists to find themselves in a situation that they have received requests from different colleagues (clients). Those requests are all demanding but can potentially compete for the available analytical development resources. There may be different priorities of the projects, but the colleagues or clients possibly all regard their work a higher priority and request the full analytical support to be allocated to their work first. In order to negotiate and seek agreement among the teams/projects, it is helpful if the analytical scientists can present the workload in a way that is quantitative and easy to view. Figure 3.13 shows two time-bars that represent the tasks and timelines for project α and project β that are managed by the same lead analytical scientist. Those two time-bars visualize the tasks that need to be done, resources arrangement, and potential overlaps of the scheduled tasks. It also shows how much buffer in time is allocated for the bench scientist to take a break, even in the midst of having to test samples for multiple projects. The lead analytical scientist should always keep in mind that human beings are not machines and can make mistakes, especially when there are lots of tasks with no breathing room.

Follow-up and follow-through are project management 101. Even if we have all the good planning, without a way to track the execution status, all our effort of planning can be a waste. To track, the fishbone diagrams can be used, too. Figure 3.14, as merely an example, shows that the tasks that are not yet completed can be put in red fonts, and the tasks that belong to WIP are in brown fonts

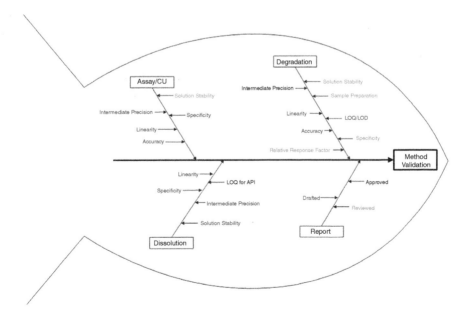

Figure 3.14 A fishbone diagram used to track the execution of the tasks for a method validation.

(note that yellow fonts are not very obvious on the computer screen or paper) and the tasks that are completed are in green fonts. The analytical scientist examines daily the fishbone diagrams and updates the task completion status in a real-time fashion. New tasks will be added whenever they appear. Since the tasks in different projects are more or less similar, the fishbone diagrams can be duplicated and modified to tailor the needs for different projects. The work status can be easily shared among the team.

3.2   Project Management – Waterfall Versus Agile

Most of the companies have dedicated project management teams. The professional PMs organize project kick-offs, plan with the team the project timelines, estimate required resources, monitor task executions, facilitate risk assessment and mitigation, ensure completion of deliverables, expedite collaboration among various functional areas, solve or help to solve conflicts among different priorities and/or between team members, and close the project after completion or cancellation. However, for a lead analytical scientist who has a variety of tasks to complete and lots of deliverables to meet, he/she has to be able to manage his/her own and his/her team's responsibilities and priorities.

There are two mainstream project management approaches. One has a "Waterfall" style and is represented by the Gantt Chart, and the other one is referred to as "Agile" and is often associated with the use of "Scrum." Both approaches are extensively described, discussed, and taught in books, on the Internet, or through training courses available for the readers to explore, and thus no comprehensive references will be listed in this book. The only reference is the book "Essential Scrum: A Practical Guide to the Most Popular Agile Process" [1].

An example Gantt chart is provided in Figure 3.15 for illustration purposes. The waterfall approach plans upfront in detail the product features, development steps, how much time is allocated for raw material purchase and release, how much time can be spent on preformulation, how many early prototypes are to be made, how much time is needed for formulation development, how many lead prototypes are to be put on stability for how long, when to conduct clinical trials or consumer studies, when to complete the regulatory filing and anticipate how long it takes for the health authority to review and approve, and of course, time and resource needed for the analytical development to support each and every step planned. The good thing about a Gantt chart is that it *provides a big picture* to the project team and also *shows the dependency between each step or each deliverable.* Each function can see how their work impacts the other functional areas and the PM or the product development team members can hold each other accountable for agreed deliverables and thus ensure the project is on track. On the other hand, it has to be kept in mind that by nature a new product development is to

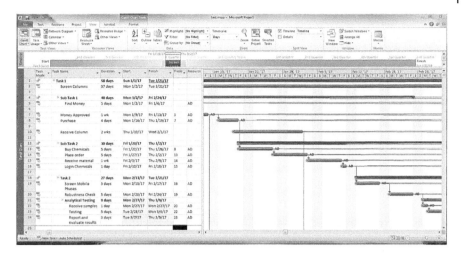

Figure 3.15   An example Gantt Chart with project activities and timelines.

develop something that is unique or unknown. It can be hard to predict what the team will encounter along the journey, especially at the project initiation stage. The initial resource estimation can be vastly off from the real consumption or spending and can cause delays, potentially as a domino effect, in completing the promised deliverables. It is time-consuming to go over the timelines with all the functional areas updating their corresponding status and therefore often the project review meeting occurs on a monthly basis. Since it involves so many functions in the timeline or deliverable review meetings, a typical one-hour meeting is not enough to cover all the topics. It is not uncommon that the team members who like to talk dominate the meetings and others sit in the meetings with their minds potentially elsewhere. The meetings may end up finishing with discussions or talks without forming concrete action items, and not rare the same topics are discussed over and over again in those meetings. A common phenomenon is that a meeting is not finished with much-needed decisions but finished by "Oops, we are running out of time, and I need to go to my next meeting."

Agile project management, in contrast, views the waterfall planning and execution wasteful and not realistic. One critical element in Agile project management is *Scrum*. It is a term borrowed from the sport of rugby. According to Wikipedia, "A scrum (short for scrummage) is a method of restarting play in rugby that involves players packing closely together with their heads down and attempting to gain possession of the ball." In rugby games, several players push forward in the same direction by forming a compact cohesive unit, with their bodies held together tightly. Similarly, a Scrum team in product development is a collaborative team that pushes forward the project deliverables based on frequent and real-time interactions and communications with great transparency. Scrum takes an iterative and incremental approach to develop products and manage the work. Figure 3.16 depicts the iterative nature of Agile project management. A team of three forms a Scrum team (the most left side cartoon with three-people-sitting-around-a-table) to work on a project. They meet and make plans for product development. Then they work independently and meet again the next day during their Daily Scrum meeting. In the Scrum meeting, they exchange updates, identify critical hurdles, get aligned on the next step, and start working independently again. Then they meet again the next day to exchange updates. After a few weeks or one month, the team conducts a Sprint review (Sprint I in Figure 3.16).

With the end goal in mind is the core spirit of Agile project management, whch is also one of the "7 habits of highly effective people." [2] To put that principle in action, Agile project management employs an important approach, called "Sprint." The project manager in Agile project management has a different name, Scrum Master. In the Information Technology (IT) world, to ensure the delivery of what the customs want at the end of development, the scrum master organizes work in iterations or cycles up to a calendar month to deliver the so-called *shippable* product increments at each predefined timeframe, i.e. Sprint. In other words, instead of delivering different parts or components of the product at each Sprint, the team delivers a product increment that resembles the final product image at each Sprint review.

The purpose of the Sprint review is to examine the shippable product that the team has produced. The shippable product has to match the image of the final product that the team designed and has the critical attributes that the company desires. After the first Sprint review, the team goes back to work and repeats the Daily Scrum – Work – Daily Scrum – Work rhythm, until the second Sprint review. Along the journey, the team holds several Sprint reviews, and in the end, they get the work done as a team.

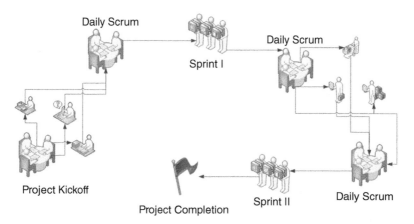

Figure 3.16 Illustration of the iterative nature of the Agile project management with Daily Scrum and Sprint Review as its core approaches.

So what are the differences between the plans, meetings, and executions following the Waterfall and those following the Agile style?

In his book [1], Kenneth S. Rubin describes what he learned from an extreme skier. The author asks whether the skier plans his entire route down when he is at the top of the mountain. The answer from the skier is no, and instead, the skier has the overall goal in mind but only plans two or three turns at the beginning, for him to reach to a certain distance. Planning detailed steps for any further distance at the initial stage is regarded as impossible and wasteful from the highly skilled skier's point of view. Indeed, the journey of new product development in a pharmaceutical R&D environment is somewhat like the ski terrain. No matter how much the Active Pharmaceutical Ingredients (APIs) are known and studied, no matter how experienced the team members are, there is always some uncertainty or uniqueness associated with developing a new product. The team or project manager can create a detailed up-front plan on a Gantt chart, with the flow and time allocated to each milestone that seems to make perfect sense on paper. It is rare, however, that a new product development project can be carried out from initiation to the finish line without any surprises. The danger of making such a detailed plan at the beginning of a project can potentially lead to a tendency that the team, the project manager, and the higher management believe the plan is correct to the point of ignoring real-time data or results. Therefore, in new product development, ***the proper balance between up-front prediction and just-in-time adaptation*** should be constantly pursued. Scrum approach is particularly well suited for operating in a complex domain. In such situations, the ability to probe (explore), sense (inspect), and respond (adapt) is critical. The Scrum approach emphasizes constant inspection, adaptation, and communication transparency. However, there is a "catch" in the Agile project management or Scrum approach. That is, a critical prerequisite for a successful Scrum team is that every team member has to be highly competent. The team members represent different functions and have to have the authority and appropriate judgment to make decisions at the spot. There has to be a high level of trust among the team members. In addition, the size of a Scrum team has to be relatively small, usually no more than six to seven people.

The traditional Waterfall project management approach works well in situations where issues are clear, or processes are well established and repetitive. One example of those situations is to plan activities according to a routine manufacturing process that has already been validated. Even the Quality by Design, to some extent, is kind of a Waterfall approach that emphasizes planning the entire activities upfront and executing the tasks based on a good design. However, when it comes to new product development, the team is in theory exploring unknowns. Especially at the beginning, there are many moving parts from technical feasibility and logistics aspects, and thus continuous planning based on the new finding becomes a necessity. Scrum approach encourages constant checking on directions, continuous replanning and keeping the end goal in mind. To achieve that,

Daily Scrum and Sprint Review are essential aspects of Agile project management. Every day, the development team members hold a short (15 minutes or less) Scrum meeting. A common approach of conducting a daily Scrum is that the Scrum Master or the team leader or the project manager asks each team member to answers three questions:

1) What did I accomplish since the last daily Scrum?
2) What do I plan to work on by the next daily Scrum?
3) What are the obstacles or impediments that are preventing me from making progress?

These three questions are typical Scrum meeting questions that the readers can find in references about Agile project management. But the Scrum Master should ask a fourth question: How can I help you? This fourth question is added by the author of this book. For a team leader or an analytical development manager, the question "How can I help you?" is a better question than "How come you did not finish what you promised to deliver?" The daily Scrum is not for problem-solving or detailed technical discussions. Those technical discussions should be held among involved team members separately. The daily Scrum also is not for a status update, rather, it is an inspection whether the team is working on the highest priority and whether the work is delivered on time, and what help is needed to make sure that the team can deliver. The daily Scrum activities minimize the potential of future finger-pointing among the team members since the communications occur daily within a small team. It is also important to make this daily Scrum a routine, make it a reoccurring event at the same time of the day, to make our minds and even our bodies get used to this rhythm. Those four questions should be asked and answered daily during that 15-minute meeting.

To translate what a shippable product means in an over-the-counter drug product development, let us take a brief look at a fictional forced degradation study conducted during an HPLC method development, for illustration purposes. We can roughly break the work into the following seven parts:

1) select experimental conditions such as the right chemicals/materials/HPLC column/ reference standards, etc.; and select the appropriate HPLC method;
2) place the materials under each stress condition for a predefined length of time;
3) pull the stressed materials and prepare analytical samples for HPLC injections;
4) set up chromatographic systems, make injections, and ensure the system suitability is met;
5) examine the chromatograms, integrate the chromatographic peaks, and quantitate the amount of the degradation occurred;
6) document the experimental procedures and observations in laboratory notebooks;
7) summarize and report the results.

The *parts* or *components* of this forced degradation study are the above seven items, and the final *product* is the outcome of the study, documented in a high-quality research report. Sprint reviews will be set regularly such as monthly, biweekly, or weekly, depending on the workload and urgency of the study. The shippable product is then a product image that represents the *successful completion of the study*, and not just, for example, *successful completion of sample stress by end of the month or week*, or *successful system suitability is achieved*. With the shippable product concept in mind, instead of planning to do things one part or one step at a time, the scrum team can plan in a way that (even if just) one sample has gone through all the steps, yes, from Step 1 through Step 7. The team thoroughly examines the executions of the work associated with each step, from stress to data summary, at the very first Sprint review. The team should allocate sufficient time for the Sprint reviews, especially the first Sprint review. If there is any defect in the shippable product, it can be discovered and corrected to reduce the carryover of the mistakes to the next round of execution of the work. This is practically very important to identify a deficiency in planning, executions, and communications. For example, for the sake of illustration, if one critical parameter, such as examination of chromatographic peak purity of the stressed samples, is not performed correctly, and is not discovered in time, one can imagine how much frustration can the team have when in the end it is found that all the stressed materials have been consumed but all the work have to be repeated because the chromatographic peak purity evaluation was not conducted correctly!

The answer to the "so what is the difference" question we posed above can be summarized below. Note the followings represent the thinking from the author of this book and not some established rules or approaches from any company's practices, nor can the following steps be found from literature. The readers are certainly encouraged to come up with more innovative approaches.

1) Plans: the Waterfall project management approach represented by Gantt Chart empathizes the establishment and alignment among the team members from the aspects of timeline and deliverables. It usually has an end goal from a time-line point of view and not necessarily from a product attributes point of view. The Agile project management approach represented by Scrum and Sprint empathizes the establishment and alignment among the team members regarding the aspects of the final product image and attributes, and with the overall timeline in front of the project team. Several small Scrum teams can be formed to work on separate parts of the product while a Scrum Master oversees the progress of each Scrum team. The plans will be adjusted whenever necessary, and the Scrum Master has to be able to represent the Scrum teams to say no, although a tough job but maybe necessary, to the higher management for those last-minute changes or nice-to-have attributes after the teams have already started the work.

2) Meetings: although it does not seem to have a direct relationship with the Waterfall style, the project meetings using Gantt Chart seem to tend to have all team members sitting in one room to discuss the progress of the project on a weekly, or biweekly or monthly basis. It is not necessarily a good way of using everyone's time. The meetings can be expensive when considering the hourly pay of the attendees. The Agile meetings, on the other hand, are frequent but short, among only the relevant Scrum team members. Small Sprint reviews can be held within one Scrum team, and then large Sprint reviews are held with a lesser frequency with all Scrum teams to present and connect their parts of the product to examine the progress by measuring against the final product image and the overall project timeline.

3) Executions: many times the Waterfall style meetings will have some team members come to the meetings and tell the team that there is no update, or "still working on it." Another not rare phenomenon of those meetings is that the same topic is discussed over and over again in many project meetings, without any real breakthrough or decisions or directions to move forward. The Agile project management, on the other hand, emphasizes the executions of the tasks. As mentioned previously, the Scrum approach emphasizes constant inspection, adaptation, and communication transparency. Without execution of the tasks, there is no inspection, nor adaptation. Each Scrum team meets daily to communicate the status of executions. The big Sprint reviews are about the inspection of the executions of each Scrum teams, and how the shippable products from each Scrum team are connected or compiled to match the image of the final product.

There are many other project management or performance evaluation tools and theories. For example, many organizations use key performance indicator (KPI) to measure the effectiveness of the company, unit, department, team, or individual to achieve business objectives. Other organizations such as Google and Intel use objectives and key results (OKR) to get alignment and engagement around measurable goals between management and teams or individuals. It is out of the scope of this book to discuss further details about many different project management principles and practices. Readers who are interested in this topic are encouraged to take training, attend courses, and self-learn through online resources.

3.3 Resource and Cost Estimations

One task a product development scientist may have to manage is the estimation of resources and costs for doing one project. Although a project manager will be ultimately responsible for managing those financial aspects of a project, they have to rely on the source information from the lead scientists on the team. To quickly provide the estimations with reasonable accuracy, some excel spreadsheet templates can be handy. Figure 3.17 is an example Excel spreadsheet for cost

Time Point (M=months)	Lot(s)	Package(s)	Condition(s)	Time Points	Total Samples
Initial	2	1	1	1	**2**
1M, 2M	2	1	1	2	**4**
3M, 6M	2	1	3	2	**12**
12M, 18M, 24M, 36M, etc.	2	1	2	4	**16**

Initial time point	Studies: initial		
Tests	Unit cost	Testing	Cost
Appearance	$ 100.00	2	$ 200.00
Assay	$ 100.00	2	$ 200.00
Degradants	$ 100.00	2	$ 200.00
Dissolution	$ 100.00	2	$ 200.00
Assay Setup	$ 100.00	2	$ 200.00
Degradation Setup	$ 100.00	2	$ 200.00
Dissolution Setup	$ 100.00	2	$ 200.00
Total Aerobic Plate Count	$ 100.00	2	$ 200.00
Total Yeast & Mold Count	$ 100.00	2	$ 200.00
Escherichia coli	$ 100.00	2	$ 200.00
Stability Initiation Charge	$ 100.00		$ 100.00
Test subtotal			**$ 2,100.00**

1 & 2 Month time point	Studies: 40/75		
Tests	Unit cost	Total activity	Cost
Appearance	$ 100.00	4	$ 400.00
Assay	$ 100.00	4	$ 400.00
Degradants	$ 100.00	4	$ 400.00
Dissolution	$ 100.00	4	$ 400.00
Assay Setup	$ 100.00	2	$ 200.00
Degradation Setup	$ 100.00	2	$ 200.00
Dissolution Setup	$ 100.00	2	$ 200.00
Storage	$ 100.00	4	$ 400.00
Test subtotal			**$ 2,600.00**

3 & 6 Month time point	Studies: 40/75, 30/65, 25/60		
Tests	Unit cost	Total activity	Cost
Appearance	$ 100.00	12	$ 1,200.00
Assay	$ 100.00	12	$ 1,200.00
Degradants	$ 100.00	12	$ 1,200.00
Dissolution	$ 100.00	12	$ 1,200.00
Assay Setup	$ 100.00	2	$ 200.00
Degradation Setup	$ 100.00	2	$ 200.00
Dissolution Setup	$ 100.00	2	$ 200.00
Storage	$ 100.00	2	$ 200.00
Test subtotal			**$ 5,600.00**

12,18, 24, 36 Month time point	Studies: 30/65, 25/60, 5 time points		
Tests	Unit cost	Total activity	Cost
Appearance	$ 100.00	16	$ 1,600.00
Assay	$ 100.00	16	$ 1,600.00
Degradants	$ 100.00	16	$ 1,600.00
Dissolution	$ 100.00	16	$ 1,600.00
Assay Setup	$ 100.00	4	$ 400.00
Degradation Setup	$ 100.00	4	$ 400.00
Dissolution Setup	$ 100.00	4	$ 400.00
Storage	$ 100.00	4	$ 400.00
Test subtotal			**$ 8,000.00**

Method Validation Estimation	$ 20,000.00
Method Transfer Estimation	$ 5,000.00

Total Cost Estimation	$ 43,300.00

Figure 3.17 Example spreadsheet for cost estimation of outsourcing a stability study.

API Name	Stress Conditions					
XYZ	Acid	Base	Heat	Oxidation	Light	Hydrolysis
Stress Detailed Conditions						
Stress Duration						
Analytical Method and column						
Data Location						
KNOWN Impurities Observed						
Main Other Observations						

Figure 3.18 Example template for summarizing outcomes of an API forced degradation study.

API Name	Excipients					
XYZ	MCC	Povidone	Mg Streate	Flavor 1	Flavor 2	Color
Stress Detailed Conditions						
Stress Duration						
Analytical Method and column						
Data Location						
KNOWN Impurities Observed						
Main Other Observations						

Figure 3.19 Example template for summarizing outcomes of an API-excipient compatibility study.

estimation of a stability program that will be outsourced and will last for 36 months. The $100 price listed in the figure is for illustration purposes and does not represent any real cases. The figure is very self-explanatory and can be used to quickly generate the stability program cost estimation. Figure 3.18 is an example template table that can be used for summarizing outcomes of an API forced degradation study. As mentioned previously, teamwork is crucial for the success of a team and the success of the scientist him/herself. Knowledge sharing is essential for successful teamwork. To summarize the study outcomes in the same format that captures the key information can facilitate knowledge sharing. Similarly, Figure 3.19 shows a template that can be used to share the study outcomes of excipient compatibility results. Figure 3.20 is a template that can be used to estimate the time and FTEs for laboratory work. This template can be helpful for a lead analytical scientist to plan daily experimental work, or gauge the workload, or to evaluate the team performance.

Tests	Hours for 1 unit sample	Hours for 2 unit samples	Hours for 5 unit samples	Hours for 10 unit samples	Hours for 20 unit samples
Appearance	0.1	0.1	0.1	0.2	0.3
Assay sample preparation	1.0	1.5	2.0	3.0	5.0
Degradation sample preparation	1.0	1.5	2.0	3.0	5.0
CU sample preparation	4.0	6.0	14.0	32.0	64.0
Dissolution preparation and profile run time	3.0	6.0	15.0	30.0	60.0
Assay HPLC instrument setup	5.0	5.0	5.0	5.0	5.0
Degradation HPLC instrument setup	5.0	5.0	5.0	5.0	5.0
Dissolution HPLC instrument setup	5.0	5.0	5.0	5.0	10.0
Instrument run time					
Data processing for Assay	1.0	1.0	1.0	2.0	2.0
Data processing for Dissolution Profile	3.0	6.0	15.0	30.0	60.0
Data processing for Dissolution Single point	1.0	1.0	1.0	2.0	2.0
Data processing for Degradation Products	3.0	6.0	15.0	30.0	60.0
Notebook documentation for Assay	0.5	0.5	0.5	1.0	2.0
Notebook documentation for Dissolution Profile	1.0	1.0	1.0	2.0	4.0
Notebook documentation for Dissolution Single point	0.5	0.5	0.5	1.0	2.0
Notebook documentation for Degradation Products	1.0	1.0	1.0	2.0	4.0
LIMS/share point data upload	0.0	0.0	0.0	0.0	0.0
Total analysis hours	35	47	83	153	290
Notebook review	3.0	3.6	6.0	9.0	15.0
Total Hours	**38**	**51**	**89**	**162**	**305**
Total Proposed Realistic Result Turn Around Hours with consideration of other responsibilities, other rojects, vacation, etc.	**53**	**71**	**125**	**227**	**427**

Figure 3.20 Example template for resource estimation.

3.4 Desired Skill Sets

As can be seen from the above section, a lead analytical scientist is spread thin to cover various aspects that require different skills. Ideally, the Analytical Development team should have various subject matter experts, and the company can put the right people in the right positions to maximize the use of people's strengths. Some financially robust companies may be able to hire many analytical scientists to split the work. The majority of the reality, however, is that many organizations are asking people to achieve more with lesser resources. Therefore, if every lead analytical scientist can grow to be a super individual, it would be ideal. This can be viewed as a dilemma or a big challenge. But it also can be viewed as a great opportunity to expand one's horizon and makes him/herself a well-rounded professional and becomes very competitive on the job market.

From the scientific aspect, the lead scientist starts the project by conducting a literature search or internal knowledge space search to understand:

1) The organic chemistry, degradation chemistry, and physical chemistry of the active pharmaceutical ingredient(s);
2) The molecular structures of the proposed excipients and their roles in the formulas,
3) The potential impurities coming with the excipients
4) The most up-to-date analytical methodologies that are published regarding the analyses of the API(s) or similar products

The scientist then determines the strategies for:

1) Providing analytical leadership in the formulation development, such as conducting analytical work in house or outsourcing the work to CRO;
2) Developing analytical methods, such as using USP monograph methods due to regulatory requirements, modifying existing methods due to high similarity between the new products and historical products, or starting with the platform methods, etc.;
3) Prioritizing prototype analyses, based on resource availability and project priorities.

From management aspects, the lead scientist works with the project team to estimate the costs and resources required for the analytical work in product development. The scientist gets alignment and constantly checks with the team to ensure the project is on track, the timeline is honored, and deliverables are met. The lead analytical scientist should be able to work confidently with other functional team members, be able to present analysis results, and be able to communicate findings.

When the formula is ready to be finalized, the lead analytical scientist needs to ensure the continuity of the storytelling of the product development without causing confusion if different analytical methods are used along the product development journey. The development moves on to the next steps such as scale-up and technology transfer to the manufacturing sites. The lead scientist will start to work with a broader team to validate and transfer the methods to various manufacturing sites or contract laboratories. A preexisting good working relationship between the analytical scientist and the receiving laboratories facilitates the handover of the developed methods. In order to have a smooth method transfer and to ensure the continuity of the data generation and result interpretation, not only does the scientist need to ensure that the method development moves seamlessly from building the knowledge space to the building of design and/or control space but also needs to polish the methods to be robust and simple enough for routine use in the QC laboratories.

The list below has the knowledge or skill sets that are desired for one lead analytical scientist. Needless to say that the sky would not fall if one does not have all the skills listed below. Working hard, having ownership of the work, and continuous learning will support analytical scientists to grow professionally and eventually find him/herself possess most of the desired skills.

1) Clear understanding of the process of product development;
2) Good organization skills; keep a big picture in mind and pay attention to details;
3) Good at time management and task management;
4) Good at project management. Being able to manage work priorities, tasks, and resources;
5) Great learning agility, good at summarization and knowledge sharing;
6) Great communication skills, dare to draw conclusions at the spot based on thorough thinking behind the scene, and good at storytelling;
7) Good understanding of the chemistry of APIs;
8) Good understanding of the chemistry of excipients;
9) Clear understanding of the common degradation chemistry;
10) Familiar with various techniques of sample preparation;
11) Proficient in various analytical techniques;
12) Solid understanding of the surface chemistry of various HPLC/UPLC columns;
13) Understanding the mechanisms of various detectors used in chromatography;
14) Expert in chromatographic data process (efficiently and correctly process data and generate accurate results, etc.);
15) Good at troubleshooting;
16) Good at converting R&D methods to QC friendly methods;

17) Good at using common computer software;
18) Good at scientific writing to create high-quality documents such as specifications, methods, protocols, reports, stability reports, justification of specifications, etc.;
19) Strong GMP mindset;
20) More than a cursory understanding of Regulatory Requirements.

3.5 Analytical Scientists and Innovation

In the Consumer Health industry, innovation means everything. It is critical to the performance and even the survival of a company. The companies are not doing medical discoveries, nor are they trying to find break-through blockbuster drug molecules to cure cancers or Alzheimer's diseases. But one Consumer Health company must be able to continuously deliver new news to the market to maintain its presence and refreshes the consumers' memories to keep using the company's products. A common practice that consumer health-care companies do is that they spend lots of money and time to reach out to the consumers, by conducting surveys, performing big data analysis, or directly talking to consumers to find out what those potential customers want. It is an important but hard task to do routinely. Ironically, since this type of study is not done very frequently due to financial and resource burdens, the acquired data/information can potentially be out-of-date after a while. In addition, the gathered information can certainly be very similar to what the competitors gather through the same approaches. Moreover, nowadays the rhythm of life is so fast that the consumers are coping with their busy schedules and not necessarily think deeply about what they really want, and more rely on the innovations from the professionals. It is more and more possible that a consumer rather asks, "Can you tell me what is good for me?" then tell the health-care professionals "This is what I want" based on their own thorough considerations.

Therefore, everyone in the company is responsible for innovation, for contributing new product ideas. It is certainly no exception for an analytical scientist.

3.5.1 Think Outside the Box

A very familiar practice, especially when it comes to innovation, is brainstorming. It is not rare to see the people normally with a serious mindset, such as R&D scientists, to dress strangely in an innovation session and are asked to brainstorm to think outside the box. What usually happens in that kind of situation, however, is that either some colleagues do not know what to think and stay quiet as a bystander or some colleagues think wildly. Instead of good ideas, what are identified might just be some better talkers.

Think outside the box is great, which often requires someone to dare to attempt to complete seemingly impossible missions or to link unrelated fields. However, it should not be used routinely, especially when we are encountering a problem, or in a survival mode. It has to be based on constant thinking and researching. There is no rule or magic crystal ball but there is something people can do:

1) Link two orthogonal, non-related items, processes, concepts, and comes up with a new idea that has a breakthrough impact on the daily lives of consumers. A broad spectrum of knowledge is a must, which can be achieved by studying another industry, learning about another culture, reading unrelated books, etc. For example, online shopping is because the IT industry grows the network from 1G to 2G to 3G and 4G, which has made the internet speed fast. Somewhere someone thought the high internet speed could be used not only for playing video games but also can be used to open online stores. The speed of the internet joined with shopping created this enormous online digital virtual shopping world that has changed the way people live.

2) Have the guts to create a market, not only follow the market trends. A great example of this is the iPhone. Microsoft did not believe that the iPhone would be a success since its high price and small market share of the first-generation Apple iPhone in 2007, but the history witnessed the opposite.

Unfortunately, Think outside the box might not match the nature of analytical scientists. An analytical scientist is very much data-driven, process-based, and detail-oriented. Without a situation, a case study, a problem, an issue, or a piece of data in front of them, it might be difficult for analytical scientists to analyze, to brainstorm.

3.5.2 Think Inside the Box

The good news is, we can also think inside the box [3]. It is, in fact, a well-known concept for innovation. Think Inside the Box means approaching problems in known ways, generate new ideas based on the tangible attributes of existing products. Many detailed descriptions regarding the approaches of thinking inside the box can be readily found online. In short, new ideas can be formed based on the following manipulation of the attributes of an existing product

1) Subtraction
2) Multiplication
3) Division
4) Task unification
5) Attribute dependency

For example, Subtraction, when presented with a colorful pediatric medicinal syrup, we can think whether the product look more attractive (to some customers) if the color is removed. There is a cluster of products, especially the pediatric products, which are advertising the feature called "Dye Free." Similarly, when we are presented with a product loaded with preservatives, we can ask what if the preservatives are removed. Again in the real world, a product line of "Preservative Free" attracts many customers who are ingredient-conscious. In a way, the self-driving car is an innovation of subtracting the critical attribute of driving, the driver, from the system. Oppositely, when we look at a regular tablet, we can visualize what if we make the tablet with two layers, each layer has a different API or functional excipients such as flavors. The result is consumers find appealing Bi-layer (or Tri-layer) tablets on the market. This example uses the Multiplication as the thinking tool. Furthermore, when we are presented with a bottle of 100 tablets, we can think what if the tablets can be packaged individually so the customers can purchase only the necessary amount and carry only a few individually packaged tablets when traveling. This is the Division approach of thinking inside the box in practice. In the United States, combination products are quite popular. People can buy medicinal products that contain multiple APIs for treating cough and cold symptoms. The thinking tool here is Task Unification. The thinking tool of Attribute Dependency can find its application in the idea of orally disintegrating tablets, the formula of which breaks the dependency of needing water to take medicines. All the above product ideas are well-known examples for illustration of what the "Think Inside the Box" is. Many great ideas can be formed based on adding, reducing, rearrange the orders of the attributes of existing products. Readers can find a lot more exciting examples from other industries to further understand the power of those thinking tools.

This ideation-derived-from-existing-product approach better suits analytical scientists than the open-end brainstorming approach does. The original form, i.e. the existing product, is presented in front of the analytical scientists, who can start analyzing the original forms and come up with new forms, i.e. ideas of new products. The hurdle of this thinking process is, however, not small. Our brains make us see what we see and get used to the existing forms of the product. In other words, after we accept something, we think this is the way it is supposed to be or supposed to function, or this is the order of attributes that are supposed to follow, then no other (better) alternatives may exist. We are limited by what we know, or what we think we know. This is so-called Cognitive Fixedness: a mindset where objects or situations are perceived in specific ways, excluding any alternative. Fixedness is the inability to realize that something known to have a particular use may also be used to perform other functions. This cognitive fixedness might, ironically, be a natural hurdle of an analytical scientist who is usually good at pattern recognition.

Therefore, conscious practice of using our analytical skills to innovate by thinking inside the box is a worthy exercise for the analytical scientists who are determined not only to deliver data but also to lead product innovation.

3.5.3 Be Analytically Creative

Whether thinking outside the box or thinking inside the box, it does not mean the ideas are only about creating new products or coming up with new claims. Be "analytically creative" means that as analytical scientists, we can participate in innovations by using the tools that we use daily. For example, in order to evaluate the organoleptic attributes of a new product, it costs a lot of money and resources to conduct real consumer studies. Moreover, the costs prohibit a survey of a large number of prototypes. However, analytical scientists can use tools such as HPLC and GC that we use daily to conduct the evaluation in the laboratory. Although it appears to be formulation scientists' job to create a product that has pleasant flavors, is less bitter, or is more comfortable to swallow, analytical scientists should not restrict themselves by any real or artificial boundaries.

Figure 3.21 presents an example diagram of a fictional experimental setup of an artificial throat and an artificial nose: measuring the tablet swallowability and the release of flavor ingredients. Figure 3.21 does not represent any actual laboratory settings in any company. Same as many proposals, figures, and templates in this book, the design is to provoke interested readers to think and come up with their own, real ideas. In Figure 3.21, the artificial saliva is pumped into a glass tube, which is connected with a rubber tube and the connection is secured with a clamp. The artificial saliva flows from the right to the left. Oppositely, carrier gas flows from the left to the right and goes out through the gas outlet. A tablet is placed after the three-way valve. To test the swallowability of the tablet, the glass tube is tilted by a foreign force in a controlled way. The time from the tablet starts moving to the time the tablet falls out of the rubber tube is measured and used to compare the swallowability of tablets with different formulas, coatings, or sizes. The carrier gas will blow the flavor that is released upon the contact of the artificial saliva and

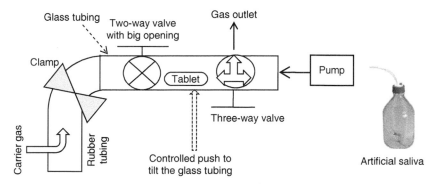

Figure 3.21 Example diagram of a fictional experimental setup of an artificial "throat" and an artificial "nose": measuring the tablet swallowability and release of flavor ingredients.

the tablet through the gas outlet, where a GC detector or a sampler can be connected to measure the strength of the flavor going to the "nose."

Figure 3.22 presents an example diagram of an experimental setup, an artificial tongue, that can be used to compare the bitterness of different formulas. Fundamentally, a bitter molecule has to be dissolved in the saliva before it can be tasted as bitter. Therefore, to evaluate the effectiveness of the taste masking, such as tablet coating or adding flavors or other excipients to compete for the saliva, the experimental setup in Figure 3.22, can be helpful. Artificial saliva (or simply pH 6.8 phosphate buffer) can be pumped through a hollow HPLC column. The column is an out-of-service column that does not offer acceptable chromatography anymore and thus the packing silica materials are removed. A crushed tablet (for mimicking chewable tablets) or a one-dose aliquot of a formula powder is placed in the empty column. The artificial saliva flows through the column and is collected in a few beakers, each at about 5–30 seconds interval. The time has to be short enough since the bitter taste is usually sensed within seconds or no more than a few minutes. The collected artificial saliva samples are analyzed by HPLC and the amount of dissolved bitter molecules can be measured. Using this approach, the formulas under development can be compared to some existing products so that the effectiveness of the bitterness masking can be assessed.

The above experimental settings are presented in this book only to make a point that the analytical scientists can actively contribute to innovations. Many other

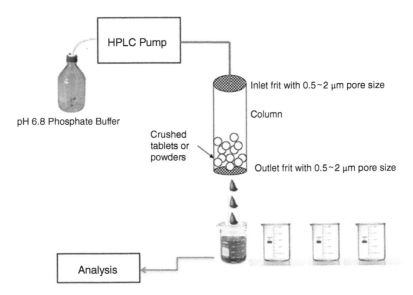

Figure 3.22 Example diagram of an experimental setup that can be used to compare bitterness of different formulas.

analytically creative approaches must be out there waiting for the analytical scientists to turn them into reality.

References

1 Rubin, K.S. (2012). *Essential Scrum: A Practical Guide to the Most Popular Agile Process*, 1e. Addison-Wesley Professional.
2 Covey, S.R. (2013). *The 7 Habits of Highly Effective People*. Simon & Schuster, 25th Anniversary Edition.
3 Boyd, D. and Goldenberg, J. (2013). *Inside the Box: A Proven System of Creativity for Breakthrough Results*. Simon & Schuster.

4

Analytical Chemistry and Separation Science at Molecular Level

Analytical chemistry studies the methodology on detection and determination of chemical species. In some cases, the signal of detection is based on direct and selective interaction between the target analyte in a sample matrix and the detecting component on the analytical equipment. The signal intensity is proportional to the mass of the target analyte within a specified sample concentration range. Examples of direct analyses include chemical sensing which produces analytical signals from electrochemical measurements such as potentiometry, amperometry, voltammetry, and coulometry; or from spectrometry determinations such as nuclear magnetic resonance (NMR), infrared (IR), ultraviolet-visible (UV-vis) spectroscopy, etc. Some analyses start with a separation (including extraction, purification) of the analyte from the sample matrix or other coexisting components, followed by subsequent detection and determination. The science studying the mechanisms regarding separation and its subsequent detection and determination is referred to as Separation Science, which has another highly popular name, that is, Chromatography.

Chromatography represents the main application of analytical chemistry in the pharmaceutical industry. Analytical method development, to a large extent, implies the development of test procedures based on reversed-phase high performance liquid chromatography (RP-HPLC). It involves selecting fit-for-purpose chromatographic conditions including an HPLC column, a mobile phase combination of salts, buffers, and organic solvents, and a robust gradient program that delivers the components of the mobile phases through the column with just the right eluting strength. In the past few decades, many novel HPLC columns with various surface modifications aiming at bringing forth unique separation features have been commercialized. With lots of different HPLC columns available, it seems challenging for an analytical scientist not to have a "choice phobia disorder." Screening HPLC columns in conjunction with screening mobile-phase

Analytical Scientists in Pharmaceutical Product Development: Task Management and Practical Knowledge, First Edition. Kangping Xiao.
© 2021 John Wiley & Sons, Inc. Published 2021 by John Wiley & Sons, Inc.

compositions and meanwhile screening mobile phase gradient variations can consume enormous time and resources of the analytical development. Nevertheless, many publications and conference presentations advocate "Analytical Method Development Quality by Design (QbD)" that is based on extensive screening of chromatographic conditions to find a perfect separation of all the known and unknown compounds in samples. In a fast-paced new product development environment, however, the key is to learn how to efficiently develop a good enough HPLC method without screening dozens of columns, gradients, or mobile phase combinations, since the extra effort may not provide enough of a benefit to justify the cost and time of such an analysis. Regardless, a critical prerequisite for any analytical method development QbD is that a valid QbD can only exist if the analytical scientist is truly a designer. The scientist should understand the chromatography at the molecular level to design the separation, and not just to screen all the possible parameters or combinations of the parameters. The design is based on the scientist's understanding of chemical interactions (1) among analyte molecules; (2) between analyte molecules and surrounding mobile phase water/solvent/salt/additive molecules; (3) between analyte molecules and column stationary phase hydrocarbon chains (with or without in-bedded functional groups) plus any functional groups (such as silanol groups) on the silica particle surface; and (4) among the column stationary phase hydrocarbon chains.

Every analytical scientist working with chromatography knows the word selectivity/specificity, and if being asked how to start a chromatographic method development, the answer might be "studying the molecules." Yes, it is essential for a separation designer to understand molecular interactions. Here, a brief introduction of a concept called *molecular recognition* can be beneficial.

Molecular recognition is not something mysterious. Molecular recognition at the biological cell surface induces signal transductions and causes changes such as active (uphill) transport of ions, membrane potential changes, and ion-channel passage. The *recognition* is a highly selective interaction between a pair of molecules such as a specific receptor molecule on the cell surface and a corresponding specific stimulating protein. To mimic the signal transduction in the biological systems, scientists have extensively studied supramolecular chemistry, which examines the molecular recognition between receptor molecules and analytes. Introduced by Emil Fischer in 1894 as the *Schlüssel-Schloss-Prinzip* (lock and key principle) [1], the concept had been referred to as host–guest chemistry, inclusion complexes, and clathrates. Molecular recognition is the latest expression. Lehn expressed molecular recognition as the selective binding of a substrate by a receptor to form a supramolecular species and called the chemistry concerned about these supramolecular species as supramolecular chemistry [2]. To chemically recognize it, the analytical reagent has to have a selective interaction with the target analyte. This recognition event can occur in a bulk aqueous solution or in an organic solvent phase, in which the information from the molecular recognition can be directly or indirectly transduced into accessible signals by some physical method such as IR, NMR, or UV-vis spectroscopy. Also, the recognition events can happen at the interfaces of two phases, which induce ion/molecule transport, distribution, or charge separation. All those interactions are then translated into electrochemical signals for analytical scientists to "see" what is happening.

Principles used in molecular recognitions involve electrostatic interactions between permanent charges, charge–dipole (i.e. ion–dipole) interactions, van der Waals forces, electron donor–acceptor interactions, and hydrogen bonding. The magnitudes of the interactions between molecules differ dramatically between what occurs in the bulk aqueous solutions or organic solvents and the interactions happening at the interfaces of two phases. This is because the environment surrounding the molecules greatly influences the magnitudes of molecular interactions and therefore impacts the success of subsequent recognition events. In the case of chromatography, the common understanding is that the separation is a result of partitioning of the analyte molecules between the HPLC column stationary phase and the bulk mobile phase. Hence, the understanding of the differences between the molecular interactions in bulk solutions and at interfaces of two immiscible phases (liquid–liquid, liquid–solid), and the understanding of how significantly the surrounding environment (i.e. the mobile phases) could impact the strength of such interactions, is the basis for an analytical scientist to design selectivity into the separation.

As a side note, although an analytical scientist may hope the HPLC column surface chemistry can be designed based on some molecular recognition principles for the specific separation tasks in hand, the reality is that the column surface chemistry is predefined by the column manufacturers. The analytical scientist can only search or select a most suitable column to get the separation job done. From that point of view, luck does play some role in method development.

One disclaimer before going to the next sections is that this book is not aiming at conducting in-depth discussions on the theories or mathematics of molecular interactions, nor will it describe in detail the chromatographic separation mechanisms or HPLC column surface chemistry. The specific information can be readily found by the readers through the internet search or by reading college textbooks. However, to understand the knowledge and digest the information thoroughly, an education background in chemistry-related field is desired. Otherwise, although it is not impossible, it requires a lot of hard work and dedication to understand the related sciences. This book tries to describe the sciences from introductory viewpoints, to connect apparently different scientific fields which may share the same underlying principles at molecular levels, to point out the direction of self-learning, to open the gate of self-study, so the readers who are interested in further learning can find a reference point to start exploring. The following sections will start with presentations of some basic concepts that are important to analytical scientists, followed by applications of the knowledge in chromatographic method development.

4.1 Ions and Ionic Strength

The strongest intermolecular forces are electrostatic attraction and repulsion between ions. This charge-based interaction is nondirectional and nonselective. Ions can be categorized as (1) elemental ions such as sodium (Na^+), potassium (K^+), chloride (Cl^-), oxygen (O^{2-}), etc.; (2) inorganic molecular ions such as perchlorate (ClO_4^-), sulfate (SO_4^{2-}), phosphate ($HPO_4^{2-}/H_2PO_4^-$), etc.; and (3) organic molecular ions such as acetate (CH_3COO^-), naproxen sodium (a nonsteroidal anti-inflammatory pain reliever, Figure 4.1) and diphenhydramine HCl (a sleep aid medicine, Figure 4.2).

When examining the magnitude of charge interactions, especially for those elemental ions, it is necessary to consider the charge density and not only the number of charge(s). Charge density refers to the ratio between charge(s) and volume (size) of the ion. With the same number of charges, such as Na^+ and K^+, as we know from the elemental periodic table, the volume (size) of the Na atom is smaller than that of the K atom, and thus the Na^+ has a higher charge density (charge/volume). This higher charge density indicates that, compared to K^+, the Na^+ interacts stronger with negative ions (anions). As a result, a neutral molecule that is abundant with electrons, i.e. with negative polarity, such as the oxygen atom on a water molecule H_2O, can interact with the (positive) Na^+ stronger than does with K^+. On the flip side, a valuable insight from this knowledge is that since the interaction between K^+ and O atom on the water molecule is weaker, the K^+ can get away from water molecules easier than Na^+ can. Translate that to chromatography language, it means K^+ may get itself out of the "cage" of the surrounding

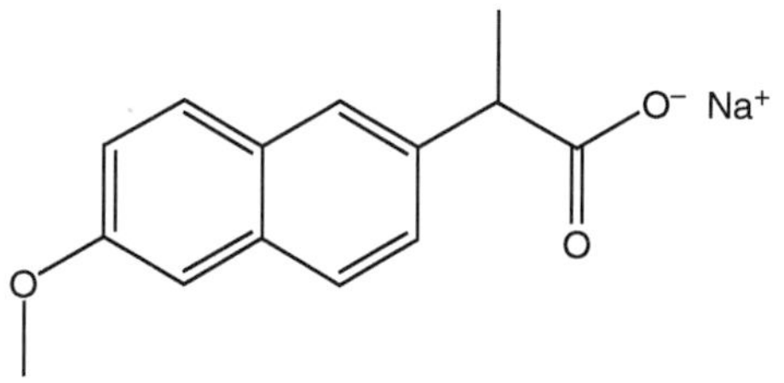

Figure 4.1 Molecular structure of naproxen sodium.

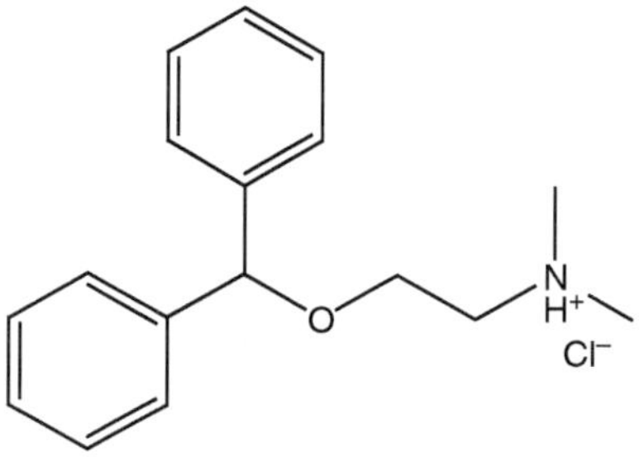

Figure 4.2 Molecular structure of diphenhydramine HCL.

water molecules in the mobile phase easier than Na^+ does. Compared to Na^+, the K^+ ion can potentially approach closer to the column surface, which is usually nonpolar or hydrophobic (e.g. C_{18} hydrocarbon chains). Understandably though, this slight difference between Na^+ and K^+ in the interactions with the surrounding environment may not be significant. However, it becomes a different story when comparing nonelemental anions such as ClO_4^- versus $H_2PO_4^-/HPO_4^{2-}/PO_4^{3-}$. The difference in the interactions between ClO_4^- and its surrounding solution environment and the interactions between phosphate ions and their surrounding solution environment can be significant. Charge density alone does not account for the difference in the case of nonelemental anions. The hydrophobicity of ClO_4^- is much larger than the other inorganic anions, and oppositely, the hydrophilicity of phosphate ions is the highest among common inorganic molecular anions [3, 4]. Indeed, when using ClO_4^- in the mobile phases, this hydrophobic anion can get itself stay at the hydrophobic surface of the column stationary phase and serves as a pseudo-ion-pair agent to help retain positively charged analytes. That is the so-called *chaotropic effect* in chromatography [5, 6]. The ability of ClO_4^- to disturb the solution environment within a very short distance from the column surface makes sodium perchlorate a useful mobile phase additive when coping with retention/separation of positively charged ions, such as amines. When adding phosphate buffers in the mobile phases, however, there is no such effect since phosphate ions show affinity toward the water molecules in the mobile phase much more than it does toward the column surface, and therefore do not come to the column surface "voluntarily."

Ions exist in solutions. Ionic strength represents the magnitude of the electric field caused by all the ions in a solution. For a salt that can dissociate into solutions as free cations and anions, the ionic strength is equal to the sum of the molalities of each cation and anion, multiplied by the square of their charges. In a simplified description, ionic strength is a measure of ion concentration in solution. The importance of the ionic strength concept is that the solubility or the dissociation constant of one salt, e.g. the ionic analyte of interest, is affected by the total ion concentration in solution, i.e. the solution ionic strength. Quite often, to obtain good peak shape for an ionic analyte, analytical scientists use salts such as Na_2SO_4 at a sufficiently high concentration to control the mobile phase ionic strength.

4.2 Protonation and Deprotonation

Whether an organic molecular ion behaves as a neutral molecule or ion depends on its protonation or deprotonation state. Protonation indicates an addition of a proton, H^+, to an atom, a molecule, or an ion. As shown in Figure 4.2, one H^+ is added to the tertiary amine moiety of the diphenhydramine molecule to form $R_1-NH^+-(CH_3)_2$. Deprotonation means an abstraction of the proton from an atom, a molecule, or an ion. As shown in Figure 4.1, the carboxylic acid moiety of the naproxen molecule changes from a neutral form of R–COOH to R–COO⁻ with Na⁺ becoming the counter cation to maintain the overall electroneutrality of the molecule.

The protonation/deprotonation chemistry is an essential concept that a pharmaceutical analytical scientist should understand. In the case of organic acids, such as the naproxen molecule, the following equilibrium in aqueous environment exists:

$$RCOOH(HA) \leftrightarrow RCOO^-(A^-) + H^+ \tag{4.1}$$

And we are all familiar with the following Henderson–Hasselbalch equation:

$$pH = pK_a + \log([A^-]/[HA]) \tag{4.2}$$

where K_a is the acid dissociation constant.

When the solution pH equals the pK_a of HA, then $\log([A^-]/[HA])$ equals zero, which means the concentration of A⁻ equals [HA]. In another word, the acid half dissociates in water, i.e. $[RCOO^-] = [RCOOH]$. When the solution pH is higher than the pK_a, the acid molecule exists more dominantly in the dissociated form, A⁻ (RCOO⁻), than the neutral form, HA (RCOOH). On the other hand, when the solution pH is lower than the pK_a, the acid molecule exists more dominantly in the neutral form, HA (RCOOH), than the dissociated form, A⁻ (RCOO⁻).

For naproxen, the pK_a is about 4. The molecule exhibits ionic property (RCOO⁻) when the solution pH is higher than 4, due to the deprotonation of RCOOH. On the contrast, more molecules exist in the neutral form, RCOOH, when the solution pH is lower than 4. When the solution pH is 1-unit away from its pK_a, the [RCOO⁻] and [RCOOH] will be 10 times different in concentrations. When the solution pH is 2-unit away from its pK_a, then the concentrations of [RCOO⁻] and [RCOOH] will be 100 times different. Therefore, if the solution pH is at or below 2, naproxen exists practically 100% as a neutral molecule, RCOOH; and if the solution pH is at or above 6, naproxen exists practically 100% as an ion with one negative charge, RCOO⁻.

In the case of organic bases, such as diphenhydramine, the protonation and deprotonation may seem slightly not that straightforward. The basicity of amine compounds, such as R–N, R–NH, R–NH₂, and R–NH₃, is expressed by the pK_a of their corresponding conjugate acids, R–NH⁺, R–NH₂⁺, R–NH₃⁺, and R–NH₄⁺, respectively. The protonation and deprotonation processes are expressed in the following equations.

$$R-NH^+ \leftrightarrow R-N + H^+ \tag{4.3}$$

$$R-NH_2^+ \leftrightarrow R-NH + H^+ \tag{4.4}$$

$$R - NH_3^+ \leftrightarrow R - NH_2 + H^+ \tag{4.5}$$

$$R - NH_4^+ \leftrightarrow R - NH_3 + H^+ \tag{4.6}$$

The conjugate acid of diphenhydramine is in the form of $(R-NH^+-(CH_3)_2)$, which has a pK_a of about 9. The molecule exhibits ionic property due to the protonation in a solution when the pH is lower than 9. The molecule exists in a more neutral form as $R-N-(CH_3)_2$ when the solution pH is higher than 9. When the solution pH is 1-unit away from the pK_a, the $[R-NH^+-(CH_3)_2]$ and $[R-N-(CH_3)_2]$ will be 10 times different; when the solution pH is 2-unit away from the pK_a, then $[R-NH^+-(CH_3)_2]$ and $[R-N-(CH_3)_2]$ will be 100 times different. Therefore, if the solution pH is at or lower than 7, diphenhydramine exists practically 100% as its conjugate acid with one positive charge due to the protonation, $[R-NH^+-(CH_3)_2]$; and if the solution pH is at or above 11, diphenhydramine exists practically 100% as a neutral molecule, $[R-N-(CH_3)_2]$.

One application of the protonation and deprotonation concept in chromatography is that it helps us understand the surface chemistry of the column stationary phases. Silanol groups on silica gel (the packing materials in columns) can exist either as $-Si-OH$ or $-Si-O^-$, depending on the mobile phase pH. Furthermore, although the pK_a of the silanol can be considered as around 4, the degree of silanol ionization depends not only on the pK_a values of silanol with different acidities but also depends on mobile phase components and pH. We should keep in mind that the pK_a and pH values are solvent dependent parameters, and then they change with the composition of the mobile phase [7, 8]. If the mobile phase brings the local pH values at the silica surface to be higher than the silanol pK_a (e.g. 1 unit greater than its pK_a), then the silanol groups are mainly in the deprotonated form, as $-Si-O^-$. The silanol sites will behave like ions and interact with other molecules based on ionic forces. On the other hand, if the mobile phase brings the local pH values at the silica surface to be lower than the silanol pK_a (1 unit lower than its pK_a), then the silanol groups are mainly in the protonated form, as $-Si-OH$. Under those conditions, the silanol groups will interact with other molecules through hydrogen bonding interaction.

The understanding of protonation and deprotonation chemistry also comes very handy when evaluating the stability of formulations. Many drug molecules are either amines or organic acids. Their stability behaviors in solutions and solid dosage forms are closely related to their protonation state.

4.3 Hydrolysis of Salts

In addition to the pH–pK_a chemistry, let us not forget that the Henderson–Hasselbalch equation does not include a very important fact, that is, the water chemistry. Salt hydrolysis occurs when the salt is formed between a weak acid and a strong base, or a weak base and a strong acid. Naproxen sodium, for example, is formed between a week acid (naproxen) and a strong base (NaOH), and therefore, the solution pH of naproxen sodium will be slightly basic due to the hydrolysis of the naproxen ion. Similarly, ammonium chloride, NH_4Cl, for example, is a salt formed between the weak base ammonia and the strong acid HCl. The solution pH of ammonium chloride will be slightly acidic due to the hydrolysis of the ammonium ion. When a salt is formed between a weak acid and a weak base, such as ammonium bicarbonate, NH_4HCO_3, the solution pH has to be evaluated on a case-by-case basis. For NH_4HCO_3, the final solution pH will be slightly basic. This is because the hydrolysis of HCO_3^-, which brings the solution pH higher than 7, is more effective than the hydrolysis of NH_4^+, which brings the solution pH lower than 7.

To be slightly more quantitative, the key to the calculation of the solution final pH due to the salt hydrolysis lies in the following equation.

$$K_w = K_a \times K_b = 1.0 \times 10^{-14} \tag{4.7}$$

where K_w is the ionization constant of water, K_a is the ionization constant for the acid, and K_b is the ionization constant for its conjugate base. For example, we can estimate the solution acidity or basicity of the ammonium bicarbonate solution. The salt dissociates in water as shown in the following equation.

$$NH_4HCO_3 \rightarrow NH_4^+ + HCO_3^- \tag{4.8}$$

The ammonium ion hydrolyzes as shown in the equation below.

$$NH_4^+ + H_2O \rightleftharpoons H_3O^+ + NH_3 \qquad K_a = 5.6 \times 10^{-10} \tag{4.9}$$

The bicarbonate ion hydrolyzes as shown in Eq. (4.10).

$$HCO_3^- + H_2O \rightleftharpoons H_2CO_3 + OH^- \qquad K_b = 2.4 \times 10^{-8} \tag{4.10}$$

From Eqs. (4.9) and (4.10), we can see that conjugate base, HCO_3^-, is more alkaline than the acid, NH_4^+, is acidic. The net result of the hydrolyses of both the acid and conjugate base is that the solution pH is larger than 7. The concentration of the acid and conjugate base can be calculated based on Eq. (4.7).

In summary, when a salt contains two ions that hydrolyze, the final pH of the solution is determined by their K_a and K_b values: if $K_a > K_b$, the solution will be slightly acidic; if $K_b > K_a$, the solution will be slightly basic. Further detailed discussions on the salt hydrolysis can be readily found on the Internet or in textbooks and therefore is skipped in this book.

4.4 Charge–Dipole and Dipole–Dipole Interaction

A dipole in a chemistry term refers to the localization of electron cloud caused by the difference in the electronegativity between different atoms within a neutral molecule. The abundance in the electron cloud results in a negative pole. The deficiency in the electron cloud results in a positive pole. The molecule as a whole is electrically neutral but possesses polar ends. A neutral molecule expressing a certain level of such polarity is called a polar molecule. The strength between the polar ends, i.e. the polarity of the molecule, is expressed by the dipole moment (μ).

When an ion approaches a neutral but polar molecule to the proximity that the charges on the ion can "sense" the electron cloud on the polar molecule, a charge–dipole interaction occurs. One obvious example of this charge–dipole interaction can be found between the elemental ions (e.g. Na^+, K^+, Cl^-, etc.) and water molecules (H_2O), as illustrated in Figures 4.3 and 4.4.

Similarly, two polar molecules can "sense" each other's polarity and take orientations that are favorable to form dipole–dipole interactions from the overall energy loss point of view. The intermolecular charge–dipole and dipole–dipole forces are much weaker than the charge interaction between ions. On the other hand, the dipole interactions are directional since the polar molecules have to take suitable orientations to facilitate the charge–dipole or dipole–dipole interaction. Therefore, whether the Gibbs free energy change (ΔG) induced by the dipole

Figure 4.3 Charge–dipole interaction between sodium ion and water.

Figure 4.4 Charge–dipole interaction between chloride ion and water.

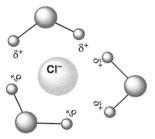

interactions is thermodynamically favorable is yet to be determined by comparing the difference between the standard enthalpy benefit (ΔH) brought to the system by the interactions and the standard entropy loss (ΔS) caused by the needed inter-molecular orientations (Eq. (4.11)).

$$\Delta G = \Delta H - T \Delta S \qquad (4.11)$$

Consequently, such dipole interactions will not automatically occur, unless the two molecules are in a vacuum. The surrounding environment in which the two molecules exist plays a significant role in terms of the overall free energy change. Since those interactions are directional, the existence of those interactions or the degree of the occurrence is different between different pairs of the polar species. Once the charge–dipole and dipole–dipole interactions occur, however, they offer a vital aspect in separation science, that is, the *selectivity*.

4.5 Hydrogen Bonding

Hydrogen bonding is a subcategory of van der Waals forces, which describes dipole–dipole interaction between molecules. Hydrogen bonds are formed between strong electronegative atoms (e.g. O, N, F, and Cl) and H atoms that are covalently bound to one of those strong electronegative atoms. The importance of hydrogen bonding is needless to say when thinking about the fact that they are responsible for the DNA structure by holding the two helical strands together. It was once believed that the hydrogen bond was quasi-covalent and that it involved the sharing of an H atom or proton between two electronegative atoms. But it is now accepted that the hydrogen bond is predominantly an electrostatic interaction. The H atom is usually not fully shared between two electronegative atoms but remains closer, and covalently bound, to its parent atom. Accordingly, the hydrogen bond between the two groups, XH and Y, is usually denoted by X-H···Y. Nevertheless, certain characteristics of hydrogen bonds do make them appear like weak covalent bonds. The O–H distance, for example, is about 0.1 nm, as expected for this covalent bond, but the O···H distance is only 0.17 nm, much less than the 0.26 nm expected from summing the two van der Waals radii but still larger than the covalent distance of 0.1 nm [9]. The electrostatic property, together with a weak covalent character, not only makes the hydrogen bonds reasonably strong but also directional. They are consequently particularly important in macromo-lecular and biological assemblies, such as in proteins, linking different segments together inside the molecules, and in nucleic acids, where they are responsible for the stability of the DNA molecules.

Hydrogen bonding plays a very critical role in the molecular recognition of many biological and artificial systems. The weak, nonspecific but directional nature of hydrogen bonds opens limitless options in the design of receptor or host molecules that include every class of organic compounds. In literature, several X-ray crystal structures show that anion binding in various biological systems results from the formation of multiple hydrogen bonds between, and with a size-selective fit of, the anion and its receptor [4]. A beautiful example of small inorganic anion recognition based on hydrogen bonding is a phosphate-binding protein (PBP) [10]. The crystal structure of PBP with bound orthophosphate was determined at high resolution (1.7 and 1.8 Å, respectively) on crystals obtained at pH 4.5 and 6.2, respectively. The electron density corresponding to the phosphate anion in the 1.7 Å structure is extraordinarily well resolved and allows the unambiguous positioning of the anion and the direct observation of the hydrogen-bonding network responsible for the anion binding. The phosphate is recognized by 12 hydrogen bonds in a cavity 8 Å from the protein surface. However, we can only feel amazed by the power of nature, which makes it seem to be so easy that 12 hydrogen bonds are remarkably obedient and work in harmony to form a molecular recognition system. On the contrary, it is just humanly impossible to synthesize a host molecule that binds a guest compound by 12 hydrogen bonds.

In 1990, Jorgensen et al., based on density calculations, have pointed out the importance of secondary contributions by neighboring atoms, which, depending on their charges, can be either attractive or repulsive [11, 12]. The ordinary hydrogen bonding can be expressed as X-H···Y, the symbol X standing for a hydrogen bond donor and Y is a hydrogen bond acceptor. Besides this primary hydrogen bonding interaction, there are also secondary hydrogen bond interactions that arise if two hydrogen-bonding-capable molecules arrange themselves, for example, with the orientation B as shown in Figure 4.5. The orientation B is more stable due to a secondary hydrogen bonding interaction occurs between O_1–H_2 and O_2–H_1. On the contrast, repulsive interaction occurs between O_1–O_2 and H_1–H_2 in orientation A due to partial positive charges on H and partial negative charges on O, and thus the orientation A is less preferred. Effective secondary hydrogen bonding interaction between different molecules sometimes can have more significant impacts on the stability or formation of a molecular complex, than merely an increase in the number of hydrogen bonds on one of the two or both of the molecules.

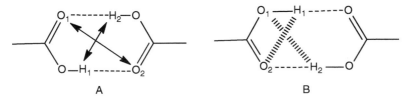

Figure 4.5 Schematic representation of secondary hydrogen bonding.

Similar to the charge–dipole, dipole–dipole interactions, the directional property of hydrogen bonding requires scientists to consider the overall change in free energy by taking into consideration of changes in both enthalpy and entropy. Philippe et al. reported the synthesis of a group of host molecules intended for selective recognition of the guest molecule phosphate. The molecular recognition is achieved by the formation of four primary hydrogen bonds between phosphate and thiourea unit on the host molecules [13–15]. The binding constant measured in dimethyl sulfoxide (DMSO) between one of the host molecules and phosphate was found to be the strongest among all known neutral host molecules that had ever been synthesized. However, that host molecule became later commercially available not because of its ability to bind with phosphate. Instead, its commercial value comes from its ability to bind strongly with chloride and has been used in chloride ion-selective electrode to measure the chloride concentration in blood ever since [16]. One learning from that study is that the binding constants between the host molecule and guest molecules such as phosphate and chloride change dramatically in different environments. When the surrounding environment is a polar organic solvent such as DMSO $((CH_3)_2S=O)$, the binding constant of the complex formed between the host molecule and phosphate is $195\,000\,M^{-1}$. In the same solvent, the binding constant of the complex formed between the same host molecule and chloride is only $1000\,M^{-1}$. A distinct selectivity of that host molecule toward phosphate seems to be achieved in DMSO. Surprisingly, the selectivity largely diminishes when the surrounding environment is changed to a nonpolar solvent, such as 1,2-dichloroethane $(CH_2Cl–CH_2Cl)$. Another host molecule which also can selectively bind with phosphate (binding constant of $820\,M^{-1}$) over chloride $(9\,M^{-1})$ in DMSO shows almost no difference in binding toward those two molecules when measured in dichloroethane. The binding constants thus measured are $1.9\times10^4\,M^{-1}$ with phosphate and $1.2\times10^4\,M^{-1}$ with chloride. The observed binding selectivity in DMSO with phosphate over chloride is most likely a solvation effect. Chloride ion interacts fairly strongly with DMSO via charge–dipole interactions and thus the binding between Cl^- and the host molecules in DMSO suffers from the competitive interaction between Cl^- and the solvent DMSO. On the other hand, due to the steric hindrance from the two methyl groups of DMSO, the interaction between bulky phosphate molecules and DMSO is relatively weak. The solvent interference to the binding between the host molecules and phosphate is presumably not as profound, and thus results in larger binding constants.

Translate the above discussions into chromatographic language is that, when hydrogen bonding interaction is designed as part of the driving force for achieving the selectivity in separation, the interactions between the analyte molecules and the column stationary phase, between the analyte molecules and the components of mobile phases, and between the column stationary phase and the mobile phases, have to be all considered.

4.6 Electron Donor–Acceptor Interaction

The interaction between molecules is induced by overlapping or repelling of the electron clouds on different molecules. In a chemistry term, a molecule with electron-rich moiety is an electron donor, and a molecule with electron-deficient moiety is regarded as the electron acceptor. For example, the H atom or proton in the hydrogen bonding is an electron acceptor, and the electronegative atom such as O, F, N, and Cl in the hydrogen bonding is the electron donor.

What makes the concept of electron donor and acceptor interesting in molecular recognition is when thinking about the interactions between two conjugated systems, such as conjugated aromatic molecules. The electron donor–acceptor interaction between aromatic rings is called π-stacking interaction [17]. Among interactions between two π systems, using benzene as a model compound for illustration purpose, the T-shaped edge-to-face arrangement, and the parallel-displaced stacking arrangement are said to be predominant. The face-to-face stacking is the least energy favorable orientation (Figure 4.6), which may be somewhat counter-intuitive to people who are not in the field of molecular recognition.

In addition, the introduction of substituents (such as Cl, F, or Br) or insertion of heteroatoms (such as N) into aromatic rings influences the relative propensities for edge-to-face versus parallel-displaced arrangement. Readers who are interested in studying in more detail of the π-stacking arrangements can find a large number of studies on this topic. However, what is relevant to analytical scientists is that we should keep in mind that this special orientation requirement of π-stacking will suffer from entropy loss. In other words, just because there are aromatic rings in structures of the analyte molecule does not necessarily mean there will be π-stacking formed between the analyte molecules and the column stationary phase that has some imbedded aromatic ring structures (e.g. phenyl columns). Whether the desired or designed π-stacking occurs or not is dependent on the overall free energy changes.

Figure 4.6 Illustration of π-stacking orientations, from left to right: edge-to-face, parallel displaced, and face-to-face arrangement.

4.7 Hydration and Solvation Energy

As mentioned in earlier sections, the environment where the molecules exist considerably impacts the strength of intermolecular forces. Take water as an example, before any solute molecules are added, the water molecules interact with each other in harmony based on hydrogen bonding. When the solute molecules infiltrate the network of water molecules, several things can happen: (1) the solute molecules start to depart from each other to interact with water molecules; (2) the network of water molecules starts to break up so the water molecules can take up the solute molecules; (3) water molecules rearrange the network which now has both the hydrogen-bonding-based water network and a new network formed based on whatever the intermolecular forces formed between the solute molecules and the water molecules. All those interactions among the water molecules, between the water molecules and the solute molecules, and among the solute molecules, result in the formation of a solution. From the analytical chemistry point of view, the process of a solution formation is a dissolving process, i.e. dissolution, which is more of a kinetic term. When thinking from a physical chemistry point of view, the aqueous dissolving process is referred to as hydration, which is more of a thermodynamic term. Figure 4.7 illustrates the events happening when a loratadine molecule tries to enter the water network. To accomodate this large loratadine molecule, the water molecules have to break their hydrogen bonding network and try to find some functional groups on loratadine to interact with this guest molecule. The pK_a of loratadine is about 4.3, so when pH is higher than 5, the molecule exists more in its neutral form than protonated form. As Figure 4.7 shows, there are not many places on the loratadine molecular structure for the water molecules to interact by hydrogen bonding. The end result is that loratadine does not dissolve well in water at pH higher than 5.

The energy change accompanied by the interaction between water molecules and solute molecules is hydration energy. For solute ions, the magnitude of hydration energy depends not only on their charge density but also depends on how

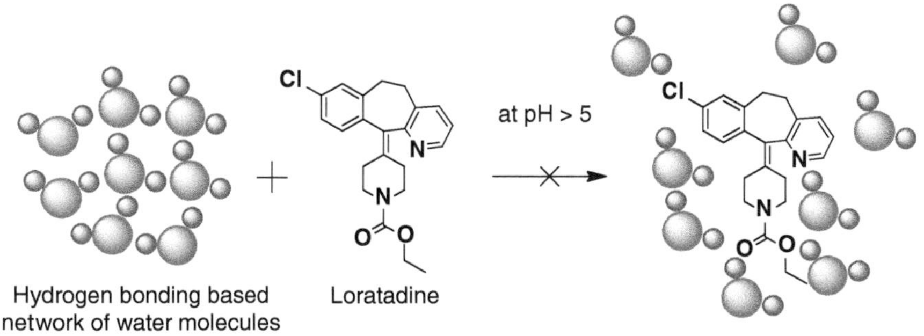

at pH > 5

Hydrogen bonding based network of water molecules

Loratadine

Figure 4.7 Interaction between water molecules and loratadine.

much hydrogen bonding those solute molecules can form with the water molecules. For example, in comparison to cations of similar size, anions have higher free energies of hydration e.g. for fluoride, $\Delta GF^- = -434.4\,kJ/mol$, while for potassium, $\Delta GK^+ = -337.2\,kJ/mol$ [3], and hence, if we want to come up with a host or receptor molecule which can selectively bind with anions in an aqueous phase, then we should take into consideration that the host molecules have to compete for the guest molecules more effectively with the surrounding water molecules. Furthermore, due to the variety of anion structures and sizes, the hydration energy of anions varies dramatically [4].

Among the common inorganic anions, ClO_4^- ion has a very low hydration energy. It is because of this low hydration energy that makes the perchlorate ion behave like a hydrophobic molecule in aqueous solutions and subsequently can create the chaotropic effect in chromatography. A chaotropic agent is a molecule in water solution that can disrupt the hydrogen bonding network among water molecules. Chaotropic solutes increase the entropy of the system by interfering with intermolecular interactions mediated by noncovalent forces such as hydrogen bonds, van der Waals forces, and hydrophobic effects. The net effect is that this ClO_4^- can penetrate through the hydrocarbon chains of the column stationary phase and modifies the environment at the very surface of the silica particles, where some leftover silanol groups exist.

Therefore, whether an interaction between solute molecules and water molecules can occur or how much it can occur, depends on the outcome of the net free energy change, ΔG. Those interactions not only break or form noncovalent bonds between molecules involved in the hydration process (an enthalpy change, ΔH) but also interrupt the water network and the solute molecule network (such as crystal lattice). The water molecules and the solute molecules may have to take certain orientations in order for the interaction to happen (an entropy change, ΔS). Moreover, differences of the hydration energies of X^- and HX are known to influence the pK_a values and thus limit the adequacy of pK_a values (in water) as a measure of hydrogen bond acceptor strengths.

When the medium is not water but an organic solvent such as methanol, ethanol, isopropanol, acetonitrile, or tetrahydrofuran (THF), the dissolving process is referred to as solvation. The energy change associated with dissolving solute molecules into an organic solvent is solvation energy. In organic solvents, the intermolecular forces can be hydrogen bonding, dipole–dipole, and electron donor–acceptor interactions. Obviously, organic solute molecules find it more energy favorable to stay in the organic solvents, in another word, the solvation energy is low for organic molecules in organic solvents.

When mixing water and organic solvents, the situation becomes much more complicated. Using a methanol–water mixture as an example, multiple networks, such as water–water, water–methanol, and methanol–methanol, will be interrupted when a solute molecule tries to dissolve into the water–methanol mixture.

The hydration energy and solvation energy are not distinguishable and thus are referred to as the solvation energy, since hydration is just one particular form of solvation. The intermolecular interactions change their type and magnitude depending on the environment. When there are more water molecules around, the formations of charge–charge or charge–dipole interactions will be more favorable. When there are more organic solvent molecules around, the hydrogen bonding, dipole–dipole, and electron donor–acceptor interactions will become dominant.

The above knowledge and mindset of thinking at the molecular level are the foundation for designing the chromatographic separations. For example, when attempting to take advantage of π-stacking effects in the separation, one should consider the changes in the hydration and solvation energy induced by the formation of designed or desired interactions that occur among the analytes, the mobile phase components, and the chromatographic column stationary phases. If we want to take advantage of some hydrogen bonding capability of a column surface chemistry, we need to know that an average single hydrogen bond between electronegative atoms can contribute up to about 30–40 kJ/mol to the binding of two partners, roughly only one-tenth the energy of a typical carbon–carbon or carbon–hydrogen single bond. In solution, however, the maximal attainable energy from host functional groups and guest analyte exclusively interacting via hydrogen bonding is severely attenuated by the dielectric permittivity of the solvent. Translated to the chromatographic language, that is, if there is too much water in the mobile phase, then the chance of taking advantage of hydrogen bonding interaction between the column stationary phase and the target analyte becomes slim.

4.8 Hydrophobic Interactions

The topic of hydrophobic interactions can serve as a good summary of the concepts we just briefly discussed, i.e. the intermolecular forces, free energy change, enthalpy change, entropy change, hydration energy, and solvation energy. The hydrophobic interaction is caused by water molecules driving the nonpolar molecules, i.e. the hydrophobic molecules, out of the water network. It is not necessarily because the intermolecular forces between hydrophobic molecules and waters are weak, nor because the intermolecular forces (mostly van der Waals forces in this case) between the hydrocarbon chains are strong. It is more related to the unfavorable loss in entropy caused by the reformation of a new water network for water to accommodate the hydrophobic hydrocarbon chains.

With that in mind, when we think about the C_8 or C_{18} hydrocarbon chains on the chromatographic column surfaces, we can conceptually imagine that the hydrocarbon chains change their spatial arrangement depending on the water content in the mobile phase. When water is the dominant component in the mobile phase, the C_8 or C_{18} hydrocarbon chains will tend to get close to each other to form "islands," or we should say the hydrocarbon chains are driven close to each other by the surrounding water molecules. With the increase in the organic content, i.e. with the increase in the hydrophobicity of the mobile phase, the C_8 or C_{18} hydrocarbon chains can spread more and do not stick to each other. Figure 4.8 illustrates the changes in hydrocarbon chain orientation in mobile phases rich in acetonitrile (CH_3–CN) (left side drawing) and mobile phase rich in water (right side drawing).

Note that Figure 4.8 is an oversimplified picture. In reality, the C_8 or C_{18} hydrocarbon chains do not stretch out nicely to the mobile phases. The chains are more tangled than stretched straight. The retention mechanism thus contains either partition or adsorption or both. Partition refers to the analyte molecules getting inside the hydrocarbon chain network, i.e. partitioning in-between the mobile phase and the stationary phase. Adsorption refers to the analyte molecules getting adsorbed on the tail side of the chain (the side that sticks to the mobile phase).

Although Figure 4.8 is oversimplified, it is, however, very useful to help analytical scientists to understand what is happening on the column surface. More detailed discussions will be provided in the next section, but we can take a quick look at the somehow counterintuitive retention behaviors of those polar-embedded column stationary phases. For example, it has been reported that the polar-embedded phases displayed a lower hydrogen bonding capacity than conventional or polar-endcapped phases and were shown to be less hydrophobic than conventional columns with ligands of identical chain length. If we remember the conceptual imagination, we just had for Figure 4.8, we know that when the high water content mobile phase is around, the hydrocarbon chains will be pushed together, and water molecules will go down near to the lower end of the

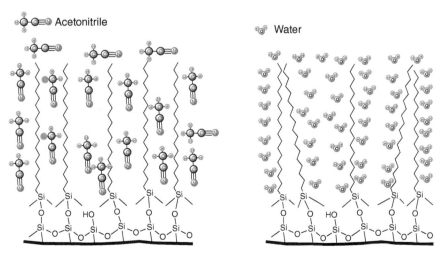

Figure 4.8 Hydrophobic interaction between C_{18} chains on a column surface changes with the amount of surrounding water.

hydrocarbon chain. The polar group, such as the amide groups on the hydrocarbon chain (Figure 4.10) will further "grab" water molecules to the silica surface and thus lowers the hydrophobicity of the stationary phase. The water retained by the amide group hinders, rather than helps, the hydrogen bonding interaction between the amide group and the hydrogen bonding acceptor analytes [18, 19]. In other words, in order to take the advantage of the potential hydrogen bonding interaction between the amide group and the hydrogen bonding acceptor analytes, a higher ratio of organic solvent maybe needed in the mobile phase composition.

4.9   Events Happening on the Column Surface

There is a lot of literature and academic research regarding the chromatographic column surface chemistry. It may make people think that the retention mechanism is very complicated. However, as far as what an analytical scientist who works in a pharmaceutical product development is concerned, the issue can be approached from a simplified angle. The retention of an analyte is to a large extent about how much the mobile phase dislikes the analytes; how much the stationary phase including the underneath silica surface likes the analytes; and how much the analyte molecules prefer to stay close together on the stationary phase.

The HPLC columns are packed with porous silica particles. On the silica particle outer surface and inside the pores, there exist countless silanol groups (Figure 4.9). Hydrocarbon chains with various chain lengths can be grafted onto

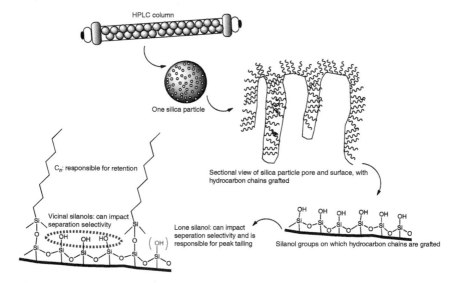

Figure 4.9   HPLC column surface at a glance.

those silanol groups through the formation of siloxane bonds ($-Si-O-Si-$). The majority of the separation takes place inside the pores rather than on the outer surface of the silica particles. One reason HPLC is a high-pressure technique is due to the fact that high pressure is needed to push the mobile phases into the pores. No matter how good a synthesis technique can be, the surface grafting of hydrocarbon chains can never occupy all the silanol groups. Not only the surface grafting technique cannot reach to a very dense level but also the space hindrance of those hydrocarbon chains will not enable one-silanol-one-chain configuration. Therefore, there will be always leftover silanol groups on the silica particle surface. Those silanol groups can significantly impact the analyte retention behaviors. Depending on the local pH of the surrounding environment, the leftover silanol groups can exist in a deprotonated form, $-Si-O^-$, or in a protonated form, $-Si-OH$, and may offer additional retention mechanisms such as charge–charge interactions, charge–dipole interactions, and hydrogen bonding. On the other hand, however, those silanol groups are generally regarded as unwanted. Endcapping is applied to those silanol groups to minimize their exposure to the analyte molecules. Figure 4.10 shows one endcapping approach that one $-Si-OH$ group is covered by one bulky trimethylsiloxy (TMS) molecule. Besides endcapping, in many cases, one C_8 or C_{18} hydrocarbon chain can be grafted on the silica surface through three siloxane bonds, which effectively covers three $-Si-OH$ groups. Various endcapping techniques, various silanol deactivation approaches, various lengths of hydrocarbon chains, various compositions of the carbon chains (such as fluorocarbon instead of hydrocarbon), and various functional groups that are imbedded or grafted onto the hydrocarbon chains (Figure 4.10) have given a large number of varieties to the chromatographic surface chemistry, and thus result in plentiful brands of columns.

Without using overcomplicated models, we can do some mental exercise by applying the knowledge of the molecular interactions we discussed in the previous sections to imagine what happens when the analyte molecules pass through an HPLC column. Let us say there is a C_8 or C_{18} column, and during column equilibration, there are lots of organic solvent molecules such as acetonitrile (CH_3CN), methanol (CH_3OH), and lesser amount of water (H_2O) molecules in the mobile phase that flow through the column (Figure 4.11).

We can (mentally) divide the space inside the column into three zones, i.e. the bulk mobile phase zone, the hydrocarbon chain stationary phase zone, and the silanol zone that is at the very surface of the silica particles (Figure 4.11). Once the equilibrium between the mobile phase and the stationary phase establishes, we can imagine that there are plenty of acetonitrile/methanol and water molecules in the bulk mobile phase. We can also safely assume that there are also a large number of organic molecules of acetonitrile/methanol that can enter the space between the C_8 or C_{18} chains. However, the very polar water molecules may

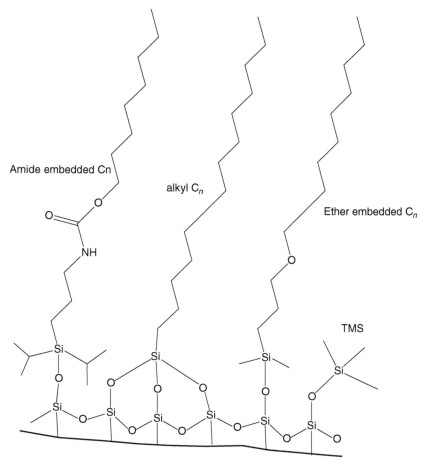

Figure 4.10 Endcapping of silanol groups; alkyl and polar (amide, or ether) embedded hydrocarbon chains.

have less desire to enter the same space between those hydrocarbon chains. Any further penetration of the water molecules down to the very surface of the silica particles, where silanol groups are accessible, maybe even more difficult. On the other hand, the acetonitrile/methanol molecules do not have much trouble to reach the surface of the silica particles.

Now imagine here comes phenylephrine, the analyte molecule (Figure 4.12).

A close look at the phenylephrine molecular structure suggests that this is an organic molecule that has a high polarity. The polarity mainly comes from the two hydroxyl groups and one amine group. To further dissect what this structural analysis means, we can think about two aspects regarding this molecule. First, as

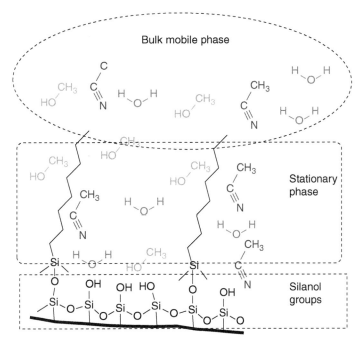

Figure 4.11 Imaginary three regions on the silica particles, (1) bulk mobile phase, (2) hydrocarbon stationary phase, and (3) silanol groups underneath the hydrocarbon chains.

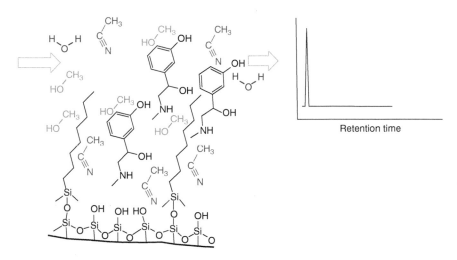

Figure 4.12 Phenylephrine molecules flow through the column, when mobile phase has a large number of organic molecules.

an organic molecule, phenylephrine likes an organic environment that is full of molecules like acetonitrile/methanol. Second, having multiple polar groups on its molecular structure makes phenylephrine like to interact with water molecules as well. Indeed, the log P (Octanol/Water Partition Coefficient) value of phenylephrine is -0.3, which shows strong hydrophilicity of this molecule. With this insight in mind, we will further imagine several chromatography scenarios. Scenario I: the amount of water in the mobile phase is minor compared to the organic acetonitrile/methanol molecules, such as 20% water : 80% acetonitrile/methanol v/v (Figure 4.12). In this case, due to the existence of a large number of organic molecules of acetonitrile/methanol, the C_8 or C_{18} hydrocarbons "freely" extend their chains to this organic solvent rich environment, similar as we see in Figure 4.8. There will be some space between the C_8–C_8 or C_{18}–C_{18} chains so that they do not pack together. However, we can predict from the structure of the hydrocarbon chains and the structure of the phenylephrine molecule, that there is not much molecular interaction can happen between phenylephrine and the C_8 or C_{18} hydrocarbon chains. It is not difficult to imagine that the phenylephrine molecules rather stay in the mobile phase, since phenylephrine likes all three molecules in the mobile phase: water, acetonitrile, and methanol. Therefore, this molecule rides with the mobile phase and flows through the column fairly quickly. As a result, the chromatographic peak elutes at very early minutes, almost at or slightly after the solvent front (Figure 4.12).

Let us now think about scenario II: the amount of water in the mobile phase is major compared to the organic acetonitrile/methanol molecules, such as 95% water : 5% acetonitrile/methanol v/v (Figure 4.13).

In this case, due to the existence of a large number of water molecules, the C_8 or C_{18} hydrocarbon chains are pushed by this highly hydrophilic environment to get close to each other to exhibit the so-called *hydrophobic interaction*, similar as illustrated in Figure 4.8. This will lead to the formation of C_8 or C_{18} chain clusters (or you can imagine them as islands in the ocean) and thus leave the space between the clusters wide open. Those spaces between hydrocarbon chain clusters are much bigger than the spaces between the hydrocarbon chains. Under that circumstance, the stationary phase zone essentially "gives up" and opens up large spaces in-between the clustered hydrocarbon chains. That then leads to the penetration of water together with the phenylephrine molecules down to the silanol zone. Once the phenylephrine molecules see the silanol groups, they interact through intermolecular forces such as charge–charge, charge–dipole, and hydrogen bonding, and the resulted chromatography is that the peak now is retained longer. However, due to the uneven distribution of silanol groups ($-Si-O^-$), and because of the strong interactions between phenylephrine and silanol groups, the resulted peak shows undesired tailing. Therefore, some of the phenylephrine molecules do not see the silanol groups and flow through the column, and some other

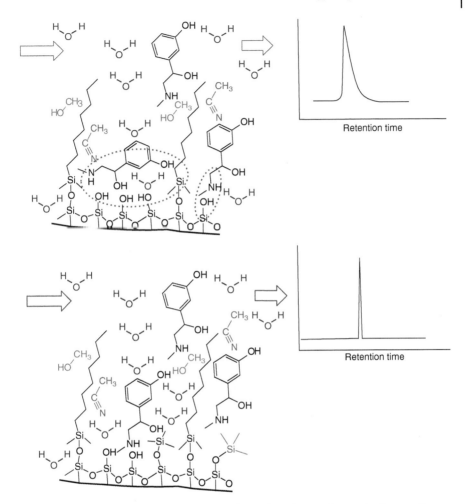

Figure 4.13 Phenylephrine molecules retain on the column, when water is more than 95% in the mobile phase. Upper figure: peak tailing occurs due to interaction between silanol and phenylephrine; lower figure: peak tailing reduces due to more endcapping of the silanol groups.

phenylephrine molecules see the silanol groups and stay longer on the column. Those microevents happening on the column surface result in macro-observation of unsymmetrical chromatographic peak shape. The right side of the phenylephrine peak drags longer before returning to the baseline. The issue can be partially resolved by selecting columns that are made of high purity silica particles with extensive endcapping (Figure 4.13).

However, although this 95% water : 5% organic solvent combination is frequently used as the mobile phase gradient starting condition to retain polar organic analytes, using a high water content mobile phase is often not sufficient to retain organic analytes that are extremely polar and small in size, such as 4-aminophenol (Figure 4.14).

In those situations, analytical scientists will either find a column that has some special surface modifications that are designed to retain such small polar molecules or apply the principle of ion-pair chromatography. Figure 4.15 shows that by adding such as hexanesulfonic acid (C_6) or pentanesulfonic acid (C_7) to the mobile phase, pairs of C_8–C_6 (C_8–C_7) or C_{18}–C_6 (C_{18}–C_7) hydrocarbon chains are formed between the column stationary phase C_8, or C_{18} hydrocarbon chains and the C_6 or C_7 hydrocarbon chains of the hexanesulfonic acid or pentanesulfonic acid, due to the hydrophobic interaction.

What happens next is that the column surface now has a layer of negative charges facing the bulk mobile phase from the $-SO_3^-$ moieties on the ion-pairing reagents. The analytical scientists now do not need to force the 4-aminophenol molecules to reach to the silica surface. Not only does this layer of $-SO_3^-$ groups offer negative charges which strongly interact with the positive charges on the amine moieties of the 4-aminophenol molecules but also enable an even distribution of those negative charges throughout the column stationary phase due to the formation of pairs of hydrocarbon chains between the C_8 or C_{18} chains and the ion-pair reagents. The end result is that the difficult-to-retain 4-aminophenol

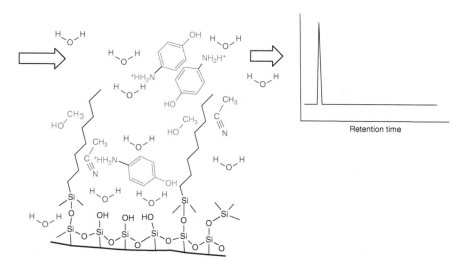

Figure 4.14 4-Aminophenol molecules do not sufficiently retain on the column even with high water content mobile phase.

molecules are successfully retained on the column by the ion-pair chromatography, and a symmetrical chromatographic peak shape is obtained (Figure 4.15).

Regarding chromatographic peak tailing, it can be easily confused with peak overloading. Figure 4.16 shows a peak shape that looks more like a right triangle and not like a right side tailed peak. This right triangle shape is very typical for column overloading [20, 21]. Although most of the literature study the overloading of basic compounds, column overloading can also happen for anionic compounds if the mobile phase pH is above the pK_a of the acidic analyte. In Figure 4.16, the example analyte is naproxen, which possesses one negative charge on its

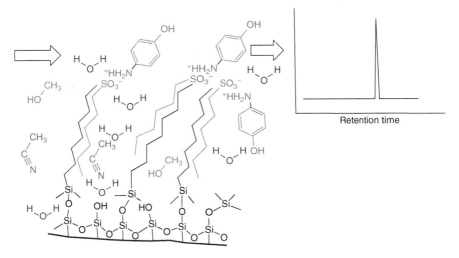

Figure 4.15 Retention of 4-aminophenol of column based on ion-pair chromatography.

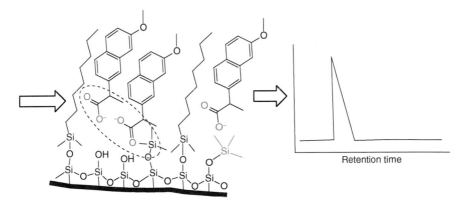

Figure 4.16 Right triangle peak shape caused by analyte overloading.

structure when the solution pH is above 6. Since there is no attraction between this negative charge and the silanol groups on the column stationary phase, the same mechanism for phenylephrine or 4-aminophenol peak tailing cannot explain the formation of such right triangle peak shape. What happens here is actually a phenomenon called column overloading caused by ion exclusion [22]. Too many of the anionic analyte molecules are retained on the column stationary phase, which results in strong electrostatic repulsion among the analytes themselves. Those analytes thus are "uncomfortable" to stay together, and preferably rushing through the column riding with the mobile phase. The resulted peak shape then reflects the "hurry" of the analyte molecules going through the column and the detector flow cell. The detector flow cell sees a "tsunami" coming, and the "wave" shoots off the roof immediately. The peak "tail" does not necessarily reflect chromatographic retention of the analyte, but most likely reflects how fast those analyte molecules can pass through the flow cell. To mitigate column overloading, increasing the ionic strength of the mobile phase can help. The existence of abundant counter ions from the mobile phase salt "eases" the repulsion among the analyte ions. The analytes can feel more comfortable to retain and move through the column in a good manner, i.e. gives a symmetrical peak shape.

4.10 Example Thought Processes of Chromatographic Method Development

4.10.1 General Considerations

Possessing knowledge and ability to apply the knowledge are two critical aspects that determine whether a professional can be successful in his/her career. In the next few sections, we are going to look at some examples of HPLC methods development by applying some of the knowledge discussed in the previous sections. The readers will find the contents and angles of viewpoints may be different from conventional chromatographic method development books or literature. The author hopes that the thought process and approaches can bring additional devices to the method developers' toolbox.

Although different method development strategies may exist, the prerequisite is that the analytical scientist is competent in science and is willing to put extra effort to get the job done. The scientist will:

1) Perform a thorough literature search for similar separations. In addition, in pharmaceutical drug product development, analytical scientists need to pay enough attention to the USP compendial methods, which are the standards for the general public. The FDA may use the USP monograph methods to test the company's commercial products. The company must do good homework during product development to thoroughly evaluate the USP methods, if available.

2) Determine the key physicochemical characteristics of the analytes, such as solution solubility, and stability, pK_a values, and UV absorbance in different solvents and/or at different solution pHs. The "Think at the Molecular Level" mindset described in the previous sections will find its application here. Everything has to be evaluated on a case-by-case basis since the molecules behave differently in different environments. Take UV absorbance as an example. It is well known that the UV spectrum of a molecule may change depending on the solvents/solutions in which the molecule is dissolved. It is not rare that the UV spectrum of a charged molecule changes its UV maximum wavelength of absorbance or changes its overall spectrum shape when the pH or solvent environment changes. For example, when an amine molecule dissolves, depending on the solution pH, the amine molecule may either be in its neutral form or protonated form. Figure 4.17 shows 4-aminophenol UV spectra obtained at pH 2.5 (left side) and 5.6 (right side). This kind of change in the UV maximum absorbance or the shape of spectra may cause inaccurate or nonrobust quantitation of the analyte if the existence of such difference is unknown to the method developer.

3) Scout combinations of stationary phases, mobile phases, and elution gradients based on educated guesses, i.e. based on experiences and knowledge. If equipped, one may use advanced technologies such as computer-assisted method development tools and automated column-switching systems to carry out stationary phase and mobile phase screening to obtain promising chromatographic conditions. As mentioned previously, the comprehensive design of experiment or QbD approaches reported in the literature may not necessarily be the most efficient and effective approach in every case especially for fast-paced product development. Of course, the opposite behavior, such as a method development is solely based on the availability of the HPLC columns in the laboratory cabinets is not appropriate, either.

4) Thoroughly examine the chromatograms and peak resolutions from the obtained separation. It is not a good practice to develop an analytical method only based on the active pharmaceutical ingredient (API) and its known related compounds. Sometimes beautiful separations can be achieved reasonably easily for those available compounds. However, a method is developed for analyzing real samples, which contains many other ingredients and unknown degradation products of the API. Stressed or aged samples must be used from the beginning of the method development. The method developer must zoom in to the baseline of the chromatograms. The scientist must thoroughly scrutinize the separation of all the peaks and preferably can remember the separation patterns, peak retention times, peak sizes, and the peak growth in chromatograms in stressed samples. The scientist should finalize the method with the purpose of the method and the potential end users in mind. The usual but unfortunate thing is that many method developers tend to babysit their

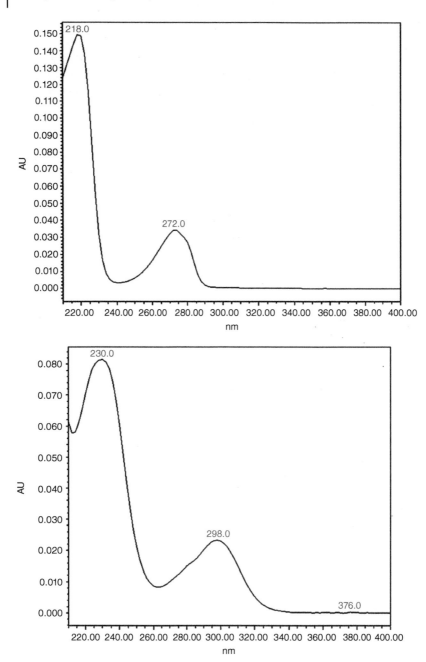

Figure 4.17 UV spectra of 4-aminophenol obtained with different mobile phase pHs: 2.5 (left) and 5.6 (right).

methods. They developed the method and therefore know the "tricks" of the method. The validation is done successfully, and the method can even be transferred successfully under the guidance of the method developer. After a while, if there is some personnel change in the method-receiving laboratory, then all of a sudden, the method does not work as well as it was used to be, because the tricks are gone with the analysts who used to run the methods routinely. This babysit mindset must be abandoned from the method development endeavor. The method developer in an R&D analytical development laboratory should keep in mind that the method is developed for other people to use. The method will be used daily in quality control (QC) laboratories that maybe even located in other countries. The method must be simple, robust, and the method procedures must be written clearly and easy to follow.

5) Develop orthogonal methods if time allows. Having an orthogonal method is sometimes necessary to ensure a thorough understanding of the degradation profile of the product. That is, however, a stretch goal since it could be time-consuming to develop an orthogonal method. Similarly, another might-be-necessary-but-could-be-time-consuming-task is that the analytical scientist may need to identify an equivalent column to ensure the business continuity, in case the primary column is discontinued or encounters some changes in its surface chemistry.

6) Be conscious of the time and resources spent. Time is always a controlling factor in the process. The skilled method developer has to balance the desire for pursuing perfection and the necessity of delivering the results in a timely fashion.

4.10.2 Case Study – Method Development for Assay of Benzalkonium Chloride

The work described in this case study was published in 2009 on the *Journal of AOAC International* [23]. The permission to reuse the material is granted by the journal publisher.

Benzalkonium chloride (BKC, structure shown in Figure 4.18) is a mixture of several homologues with carbon chain length usually from C-10 to C-16. As per

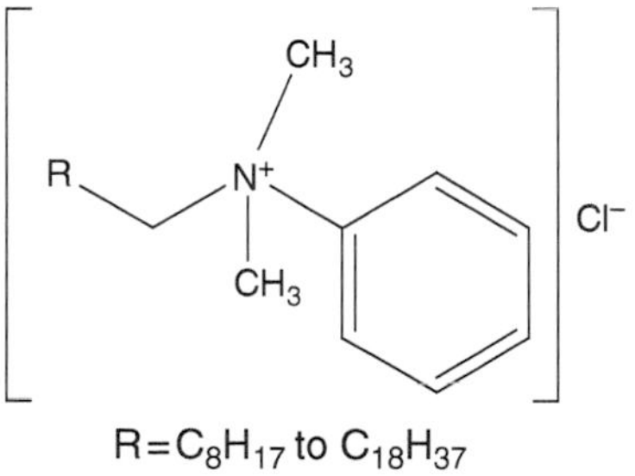

Figure 4.18 Molecular structure of benzalkonium chloride [23]. *Source:* Reproduced with permission of AOAC International.

$R = C_8H_{17}$ to $C_{18}H_{37}$

the USP requirement, the percentage of each BKC homologue, in addition to total BKC content, should be determined. The C-12 homologue must comprise at least 40%, and the C-14 homologue must be at least 20% of the total BKC content. Moreover, those two homologues together must comprise not less than 70% of the total content.

As we discussed in the previous chapter, it is the net free energy change resulted from the overall interaction between the analyte molecules and the surrounding environment that determines the chromatographic behavior of the analytes. Focusing on only one aspect of the interaction and overlooking other relevant factors will result in un-optimized chromatographic methods. A literature search revealed that most of the reported HPLC methods were mainly developed based on the consideration of the ionic nature of BKC. Chromatographic conditions, such as mobile phase pHs and ionic strengths, are the main controlling parameters in the methods. For example, the USP method condition described a mobile phase consisting of acetonitrile : sodium acetate 100 mM (55 : 45, v/v) adjusted with glacial acetic acid to pH 5.0. Similar to many other literature methods, the USP monograph method employs HPLC columns containing packing L10, i.e. HPLC columns with a cyano group (–CN) containing stationary phases. The rationale behind such a choice is presumably based on the consideration that the electron lone pair on the nitrogen atom of the cyano group (C≡N:) could interact with the ammonium group on the BKC molecules, a charge–dipole interaction in action. However, as mentioned in the general considerations section, the analytical methods must be developed with the end users in mind. The methods intended for QC laboratories must be robust and can last for a long time to be economically efficient. The retention times of the BKC homologues will, however, shift with the aging of HPLC columns having –CN group as the functional groups on the stationary phases. Under either acidic or basic conditions, the hydrolysis of –CN group forms carboxylic acid (–COOH) group. This hydrolysis changes the functional group on the column stationary phase from a neutral molecule with a –CN polar end to the negatively charged carboxylate (–COO$^-$), when the mobile phase pH is above the pK_a of –COOH, which is about 4. The interaction between the column stationary phase and BKC then changes from a 100% charge–dipole interaction between the cyano group and the BKC ammonium moiety to a mixed mode, which also includes a strong ionic interaction between the negative charge on –COO$^-$ and the positive charge on –R$_4$N$^+$. This microenvironment change leads to the increase in the retention times of the BKC homologues, i.e. an undesired retention time shift occurs.

Besides having the hydrophilic heads (the ammonium moiety), the BKC homologues are amphiphilic molecules (i.e. surfactants), which also possess hydrophobic tails (the C-10, C-12, C-14, and C-16 hydrocarbon chains). The potential strong hydrophobic interaction between the long alkyl chains of the BKC homologues and the C$_8$ or C$_{18}$ hydrocarbon chains on column stationary phases, should be taken into considerations when designing the chromatographic conditions. Interestingly, the literature search revealed that HPLC columns with C_{18} stationary phases were another popular choice for the analysis of BKC homologues. C_{18} stationary phase is not a straightforward choice if considering the intermolecular forces. We can expect that with increased length of the hydrocarbon chains of the column stationary phase, the hydrophobic interaction between the alkyl chains of the BKC molecules and the hydrocarbon chains on the column surface increases and therefore results in prolonged retentions of the BKC homologues. Along with the long retention, severe tailing of BKC homologue peaks can be anticipated. It was not surprising that the typical run times for BKC HPLC analysis ranged from 20 to 30 minutes, and the peak tailings were usually unacceptable for the homologues with longer alkyl chains.

4.10.2.1 Method Development

Now let us re-examine the events happening on the column at the molecular level. The potential interaction between BKC and reversed-phase HPLC column stationary phases can be both ionic and hydrophobic in nature. The ionic interaction can occur between the positively charged ammonium ions of the BKC molecules and the deprotonated silanol groups ($-Si-O^-$) on the column silica surface. The hydrophobic interaction occurs between the alkyl chains of the BKC molecules and the C_n ($n = 1, 4, 8, 18$) hydrocarbon chains of the column stationary phase. A closer examination reveals that the ionic interaction should not be the dominant retention mechanism for BKC homologues on the columns. The logics are (1) the charge density of BKC molecules should be small due to its bulky structure and (2) the highly acidified silanol groups are minimized with nowadays technology with the use of ultra-pure silica packing materials. On the other hand, the hydrophobic interaction is more likely to be responsible for the retention of BKC molecules under reversed-phase conditions. Figure 4.19 shows the overlay chromatograms of BKC homologues (homologue C-18 was included for concept proofing) obtained on 5 cm × 4.6 mm I.D. ACE C_4, C_8, and C_{18} columns under an isocratic condition (mobile phase consisted of a mixture of 20 mM potassium chloride solution and acetonitrile at 45 : 55, v/v). The chromatograms indicate that the retention of the BKC homologues became longer, and the peaks became broader with increased stationary phase hydrocarbon chain lengths.

Due to the ionic nature of the BKC homologues, ionic interaction can also occur among the ammonium ions of the BKC molecules and the charged molecules added in the mobile phases, such as buffers or acids. The mobile phase ionic strength has to be controlled. On the other hand, as mentioned previously, the charge density of the bulky BKC molecules should not be large. Therefore, supposedly the salt concentration in the mobile phase does not need to be very high. Different salts, including phosphate, sodium sulfate, ammonium acetate, and potassium chloride were tested. Indeed, the experimental results suggested that a

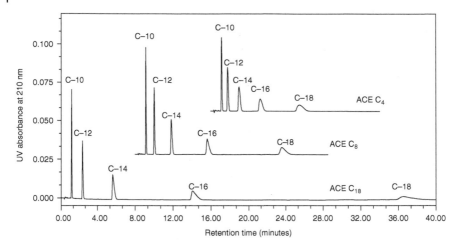

Figure 4.19 Overlay chromatograms of benzalkonium chloride homologues (C-10, C-12, C-14, C-16, and C-18) obtained on 5 cm × 4.6 mm I.D. ACE C_4, C_8, and C_{18} columns. For visual clarity, the overlay is tilted 30° from bottom to top [23]. *Source:* Reproduced with permission of AOAC International.

20 mM potassium chloride in the mobile phase is adequate to obtain stable retention of BKC homologues. The experimental results also indicate that the peak symmetry is acceptable as long as the salt concentration is higher than 10 mM.

From the literature, it seems that almost all the HPLC methods for BKC analysis control the mobile phase pH using buffer systems. However, the mobile phase pH should not affect the ionization of the BKC molecules. As ammonium ions, BKC homologues are always positively charged under all pHs. To examine whether the BKC retention time shifts when using unbuffered mobile phases, mobile phases consisting of 20 mM potassium chloride and 5 mM ammonium acetate were prepared. The mobile phase pHs were adjusted with acidic acid to be 6.8 and 5.5. The pH range was selected because an unbuffered aqueous solution should show a pH approximately ranging from 6 to 7, depending on the airborne carbon dioxide concentration in the laboratory. Although the retention times of BKC homologues increased slightly when the higher pH mobile phase was used, the separation pattern remains the same. Therefore, the requirement of a tight control of the mobile phase pH is really not necessary.

During the method development, both isocratic and gradient programs were developed that could separate the BKC C-10, C-12, C-14, and C-16 homologues within five minutes. The gradient instead of the isocratic elution was selected since the peak symmetry was better when gradient was applied. Due to the existence of some impurity peaks that eluted immediately in front of the C-12 peak in

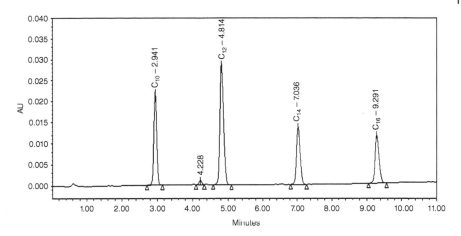

Figure 4.20 Representative chromatogram of benzalkonium chloride homologues obtained at concentrations of 0.01 mg/mL for C-10, C-14, and C-16 homologues and 0.02 mg/mL for C-12 homologue. Chromatographic conditions: 5 cm × 4.6 mm I.D. ACE C$_4$ column [23]. *Source:* Reproduced with permission of AOAC International.

the reference standard solution that was used (see Figure 4.20), the final gradient was adjusted to be 11 minutes.

At the end of the method development, a simple, rapid, robust, and stability-indicating reversed-phase HPLC method was developed for determination of BKC homologues in BKC sample solutions. Analysts from five different laboratories located in different countries successfully validated the method. The method was demonstrated to have good accuracy, linearity, precision, reproducibility, repeatability, specificity, and robustness, and is suitable for routine analysis of BKC sample solutions.

4.10.3 Case Study – Method Development for Analysis of Stereoisomers

The thought process, research work performed and literature citations described in this case study can be found in several papers published previously [24–28].

Figures 4.21 and 4.22 show the molecular structures of some representative stereoisomers. A reversed-phase HPLC separation of stereoisomers of an API can be extremely challenging to achieve. It becomes even more challenging when a baseline resolution is the goal of separation between an isomeric impurity and an API, and this isomeric impurity exists at a level of less than 0.1% of the API peak. However, it is critical to monitor isomer impurities because different isomeric forms of an API may have vastly different physiological effects. One isomer can be

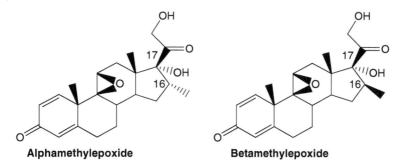

Figure 4.21 Chemical structures of betamethylepoxide and alphamethylepoxide [24]. *Source:* Reproduced with permission of Elsevier.

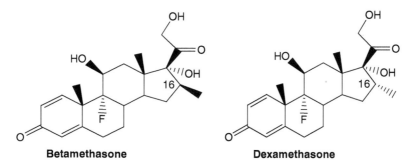

Figure 4.22 Chemical structures of betamethasone and dexamethasone [25]. *Source:* Reproduced with permission of Oxford University Press.

beneficial, while the other isomer might be toxic to human beings. Therefore, it is frequently a requirement that the API of a pharmaceutical product is in one pure form instead of existing as mixed isomers.

Before the work described in this case study was published, people had attempted the separation of betamethasone and dexamethasone by both normal-phase and reversed-phase HPLC. In the case of normal-phase chromatography, the separation was based on the off-line derivatization of the compounds. Under reversed-phase conditions, only partial separation of the two isomers was obtained with resolutions of approximately 0.9–1.5. The situation was even worse for the isomer pair of betamethylepoxide and alphamethylepoxide. No HPLC separation of betamethylepoxide and alphamethylepoxide had ever been reported. Yet those two isomeric epoxides are the key intermediates for synthesizing various steroid APIs such as betamethasone and dexamethasone, betamethasone- and dexamethasone-21-acetate, betamethasone- and dexamethasone-21-phosphate, betamethasone- and dexamethasone-21-dipropionate, etc.

From the chemical structures, it appears that none of the molecules of beta-methylepoxide, alphamethylepoxide, betamethasone, or dexamethasone have any functional groups that are easily ionizable. Unlike the chromatography of the BKC homologues, mobile phase pH or ionic strength would not affect the retention/separation of those isomeric neutral molecules. Since the physicochemical characteristics of those isomer pairs are very similar, it would be difficult for the analytical scientists to find adequate thermodynamic differences in the interaction outcomes (such as changes in entropy and enthalpy) among the epimers and the surrounding environment (the mobile phases and stationary phases). However, although the beta- and dexa-forms of these molecules are epimers with almost identical chemical structures, the orientation of the methyl group at the C-16 position is in the opposite direction from the plane. The hydroxyl group at the C-17 position may potentially "sense" the difference in affinities from hydrogen bonding acceptor in the mobile phase and/or on the column surface.

4.10.3.1 Considerations of Column Stationary Phases

Identification of the most appropriate stationary phase (i.e. the HPLC column) is critical for a successful method development. This step can be the most challenging and time-consuming as there are so many commercial HPLC columns available. Column selection should be based on the knowledge, i.e. the physicochemical characteristics of the sample (especially the key analytes) and the column stationary phase (such as type of bonded phase, bonding type, polarity or functionality, endcapping, carbon loading, hydrophobicity, particle shape, particle size, surface area, ligand density, pore size, hydrogen bonding capacity, trace metals in silica that is used in preparation of the bonded phase, etc.). Although it has been realized that the separation of isomers is possible under reversed-phase chromatography, the exact mechanism is still unknown. One thing that seems clear is that the separation of isomers cannot be achieved only by the hydrophobic interaction between the isomer analytes and the stationary phase C_8 or C_{18} hydrocarbon chains. Surface modification of the stationary phases may play an important role. For example, Snyder et al. have pointed out that C_{18} columns made from polyfunctional silanes are more effective in the isomer separations than columns made from monofunctional silane. The subtle difference in the hydrogen bonding, dipole–dipole, or other polar and nonpolar interactions between the isomers and the stationary phase surfaces, induced by the different stereo orientation of the isomers, must be responsible for the different retention behavior of the isomer molecules. Besides the properties of the stationary phases, column quality such as consistent particle size distribution, lot-to-lot reproducibility, and stability of column bed (column lifetime) should also be considered. Based on the Van Deemter equation, column efficiency increases with a decrease in the particle sizes. The reduced particle size, together with a tight particle size distribution, enhances both the interparticle and intraparticle mass transfer by shortening the diffusion distances. Improved mass transfer enhances peak efficiency and peak symmetry, and also allows separations to occur faster, thus shorter columns, higher mobile phase flow rates, and faster gradients can be used without sacrificing resolution and selectivity. In this case study, HPLC columns packed with 2 or 3 μm silica particles were selected.

Based on aforementioned considerations, a few HPLC columns were selected for initial column screening. These columns included YMC-Pack Pro C_{18}, YMC-Pack CN, YMC-Pack Phenyl, YMC Basic, YMC Hydrosphere C_{18}, TSK Super ODS, TSK Super Octyl, Ace C_8, Ace C_{18}, Ace C_{18} (300 Å), Waters Atlantis dC_{18}, Waters XTerra MS C_{18}, Thermo Fluophase PFP, and Thermo Fluophase RP C_{18}. The columns were obtained from well-established, highly reputable HPLC column manufacturers. Most of the columns are packed with the third generation ultra-pure silica particles. The selected columns contain different carbon chain lengths, various carbon loading, and diverse surface functionality. The YMC-Pack Pro C_{18} column possesses a unique endcapping procedure utilizing Lewis acid–Lewis base chemistry. The YMC Hydrosphere C_{18} column can be used under 100% aqueous conditions for separation of polar compounds, which indicating a strong potential of hydrogen bonding interaction with the analytes. Waters Atlantis dC_{18} is also good for separation of polar compounds under highly aqueous mobile phases. TSK columns are packed with 2 μm particles. ACE column surfaces are base-deactivated and are well known for offering good peak shapes. The Hybrid Particle Technology used in Waters XTerra MS C_{18} column provides an evenly distributed hydrophobicity throughout the particle backbone. Columns with Phenyl group and Cyano group imbedded stationary phases were screened because they can provide quite different selectivity from conventional C_8 or C_{18} columns. A wide-pore ACE C_{18} column with a pore size of 300 Å was also selected for initial column screening. Although wide-pore silica particles are usually used for large molecules such as proteins and nucleic acids analysis, they can also enhance the access of small molecules to the intraparticle surfaces by allowing the steroid molecules, which are not that small, to more freely diffuse into and out of the pores. In addition to the above conventional stationary phases, Thermo Fluophase PFP and Thermo Fluophase RP C_{18} were tested. The greater dipole of carbon–fluorine bond versus the carbon–hydrogen bond makes the perfluorinated stationary phase unique in the retention of polar and halogenated compounds. The perfluorinated phases have also been shown to have shape selectivity for positional/geometric isomers.

Among the above-tested columns, we put hope on the following columns: (1) YMC Hydrosphere C_{18} column due to its potential ability to bring hydrogen bonding interaction into the system and (2) ACE C_8 column due to its shorter chain length and therefore may allow the isomers reach to the column surface easier and interact through hydrogen bonding with the silanol groups.

4.10.3.2 Considerations of Mobile Phase Compositions

The commonly used organic solvents in reversed-phase HPLCs are acetonitrile, methanol, and to a lesser extent, isopropanol, and THF. To develop a QC laboratory friendly HPLC method, however, the use of THF is not preferred. Fresh HPLC grade THF has to be used to avoid high background absorbance from the trace UV absorbing impurities that are generated by the residual peroxides in aged THF (peroxides are also fire and explosion hazards). THF can also attack plastic parts, such as seals, ferrules, tubing, or filters of the HPLC systems by extracting protective additives from these plastics. The search thus focused on finding appropriate combination(s) between acetonitrile, methanol, and isopropanol that would be used as the organic modifiers in the mobile phase. The purpose of using methanol or isopropanol was again to introduce some hydrogen bonding interactions to the separation system. The target resolution factor is about 3 between the isomer peaks. One thing to point out is that for an HPLC method to quantitate isomer pairs when their concentration ratio is 1000 : 1 or more, a resolution factor significantly greater than 2.0 is needed. Although many textbooks will claim that a resolution of 2.0 is good enough for a robust HPLC separation, this is actually not that simple and straight forward. The resolution of 2.0 is usually not adequate for a true baseline separation between two peaks that differ dramatically in their peak sizes, especially when a small peak elutes after a large peak. Other factors, such as peak tailing, peak width, and different band broadening characteristics of the two peaks, would also significantly impact the baseline resolution. A resolution of 2.0 for baseline resolution works well only when the chromatographic behavior of both peaks on the stationary phase is similar, and their concentration in the sample is close to 1 : 1.

As anticipated, a mobile phase system of water and 1 : 1 volume ratio of acetonitrile and isopropanol, together with the YMC Hydrosphere C_{18} column, was capable of separating betamethylepoxide and alphamethylepoxide with a resolution greater than 3.0 (Figure 4.23). Interestingly, under the same mobile phase condition, many other columns were found to be able to separate this pair of isomers with resolution factors larger than 2. The fact that many columns were able to easily separate the betamethylepoxide and alphamethylepoxide under the selected mobile phase conditions made it look strange that how come there was no successful separation achieved before this case study. It could be, possibly, that the thought process used in this method development was truly based on the molecular structure analysis. The outcome of that thought process was the use of isopropanol, which is not always the first choice, as one of the main components of the mobile phase. As a refresh of memory, isopropanol was tried aiming at taking advantage of hydrogen bonding interaction between isopropanol and the hydroxyl groups on the steroid isomers.

It was a bit more challenging in the case of attempting a baseline separation between betamethasone and dexamethasone. The best column as predicted was

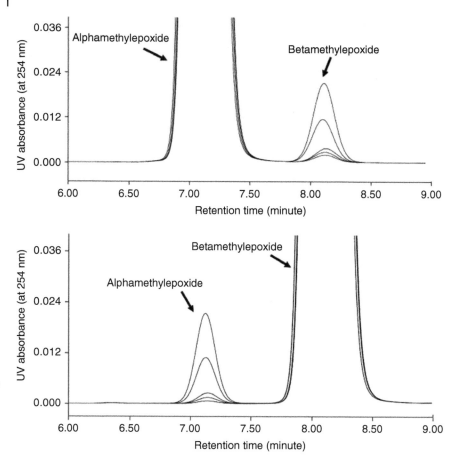

Figure 4.23 Overlaid chromatograms obtained from betamethylepoxide linearity study (top) and from the alphamethylepoxide linearity study (bottom). Only the chromatograms obtained in the linearity range from 0.01 to 1% are overlaid for visual clarity [24]. *Source:* Reproduced with permission of Elsevier.

the ACE C_8 column, but this time, the neat acetonitrile was found to be the best mobile phase organic solvent. The resolution factor between betamethasone and dexamethasone was 2.7 (Figures 4.24 and 4.25). Although it would be too much speculation, the change from epoxide to hydroxyl at the C-11 position might have interfered/competed with the hydrogen-bonding-based molecular recognition between the C-17 hydroxyl of the isomers and the mobile phase solvents, or the functional groups on the column surfaces. In other words, the increased number of hydrogen bonding interaction sites on the steroid isomers reduced the ability of the "alcoholic" mobile phase to tell the difference between betamethasone and dexamethasone.

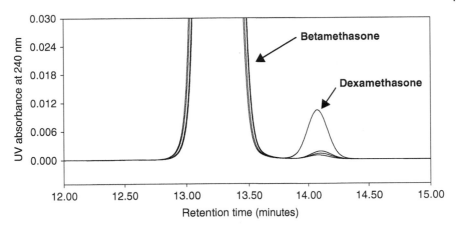

Figure 4.24 Overlaid chromatograms obtained from dexamethasone linearity study. Only the chromatograms obtained in the linearity range from 0.01 to 1% dexamethasone are overlaid for visual clarity [25]. *Source:* Reproduced with permission of Oxford University Press.

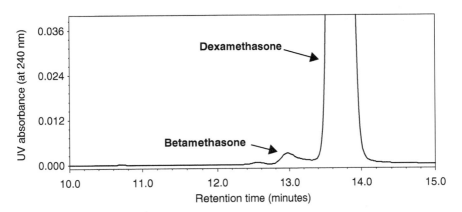

Figure 4.25 A chromatogram of dexamethasone API obtained at a concentration of 1.0 mg/mL. Betamethasone was at approximately 0.1% level as an existing impurity in the dexamethasone API. The mobile phase consisted of water and acetonitrile [25]. *Source:* Reproduced with permission of Oxford University Press.

To improve the resolution between betamethasone and dexamethasone on the ACE C_8 column, we added β-cyclodextrin to the mobile phase, and the resolution factor between these two epimers became 3.3. Cyclodextrins are well known as inclusion-complexing agents for both small and large molecules. The interior of the macrocyclic structure is hydrophobic, while the exterior is water compatible due to the existence of many hydroxyl groups. Among the commonly available native cyclodextrins, i.e. the α-, β-, and γ-cyclodextrins, the β-cyclodextrin, and γ-cyclodextrin have been extensively used as mobile phase additives in chromatography as chiral selectors for separation of various isomers including structural, diastereomeric, and enantiomeric molecules. The interactions between steroids and various cyclodextrins have been studied in detail by many research groups. The most popular hypothesis on the mechanism of separation in the presence of cyclodextrin is that the inclusion occurs primarily at the A- and B-rings of the steroids, which determines the binding strength of the inclusion complex. The hydrogen bonding interaction between the secondary hydroxyl groups of the cyclodextrins and the hydroxyl groups of the steroid molecules can create the binding selectivity between the steroid isomers. For example, the γ-cyclodextrin has been shown to less selectively bind betamethasone or dexamethasone than β-cyclodextrin does, presumably due to the larger diameter of the secondary hydroxyl rim which is 6.5 Å for β-cyclodextrin and 8.3 Å for γ-cyclodextrin, respectively. The smaller diameter of the β-cyclodextrin restricts the entry of the C- and D-rings of the steroids into the cyclodextrin cavity more than γ-cyclodextrin does, and thus provides a higher probability of hydrogen bonding interaction between its secondary hydroxyl groups and the hydroxyl group at the 17-position of the D-ring of the steroids. This hydrogen bonding interaction can be affected by the orientation of the methyl group at the 16-position, and therefore, the complexation can be different between β-cyclodextrin-betamethasone and β-cyclodextrin-dexamethasone. Indeed, the apparent association constants (K_f) of β-cyclodextrin-betamethasone complex and β-cyclodextrin-dexamethasone complex are reported as 27 and 22, respectively, in an acetonitrile–water mixture (35 : 65, v/v), while the K_f values for γ-cyclodextrin-betamethasone complex and γ-cyclodextrin-dexamethasone complex are 212 and 215, respectively. The absolute K_f values of the β-cyclodextrin-steroid complexes are much smaller than those of γ-cyclodextrin-steroid complexes, which indicate a weaker interaction due to the size restriction. The relative difference of the binding, however, is around 19% in the complexes formed between β-cyclodextrin-betamethasone and β-cyclodextrin-dexamethasone, while only 1% in the complexes formed between γ-cyclodextrin-betamethasone and γ-cyclodextrin-dexamethasone. In another word, the binding selectivity is greater in β-cyclodextrin-steroid complexes.

Separations are achieved when the mobile phases consist of a mixture of 2.5, 5, 10, or 20 mM β-cyclodextrin aqueous solution, respectively, and acetonitrile at a volume ratio of 85 : 15. As expected, the addition of β-cyclodextrin indeed greatly improved the separation between betamethasone and dexamethasone. The retention time of betamethasone and dexamethasone and the separation resolution between these two isomers were largely affected by the β-cyclodextrin concentration. The resolution factor was 3.3 when the concentration of β-cyclodextrin was 20 mM, while the resolution factor reduced to 2.4 when the β-cyclodextrin concentration was 2.5 mM. The retention times became shorter with higher β-cyclodextrin concentrations. These observations are in consistent with literature reports that the increase in β-cyclodextrin concentration would significantly decrease the retention of the analytes and meanwhile improve the separation resolution. The complexation occurs in the mobile phase, and thus due to much less retention of β-cyclodextrin, the resulted complexes move through the column faster than the noncomplexed isomers do.

A third method was developed on the ACE C_8 column by using water as the mobile phase A and THF as the mobile phase B. The resolution factor between betamethasone and dexamethasone was 3.1. As mentioned previously, THF should not be the first choice. However, THF should be tried as a last resort for a difficult separation because it may provide some interesting results due to its unique properties. The results of the trials indicated that the water–THF mixture was also effective for the separation of betamethasone and dexamethasone (Figure 4.26). The method is able to accurately quantitate Betamethasone in the presence of dexamethasone API, whose concentration was 2000 times higher.

4.10.3.3 Degradation Analysis Method Development

Once separation of the critical pair, such as betamethylepoxide and alphamethylepoxide, has been achieved, the method development can move on to the separation of other related compounds. Both betamethylepoxide and alphamethylepoxide raw materials contain many structurally similar impurities (Figures 4.27 and 4.28). Those impurities can potentially be carried over throughout the subsequent

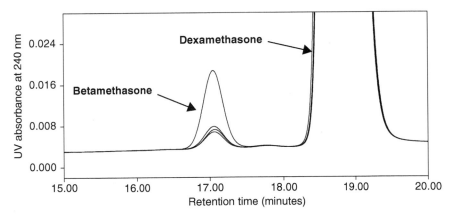

Figure 4.26 Overlaid chromatograms obtained from betamethasone linearity study, with mobile phase containing tetrahydrofuran. Only the chromatograms obtained in the linearity range from 0.01 to 1% betamethasone are overlaid for visual clarity [25]. *Source:* Reproduced with permission of Oxford University Press.

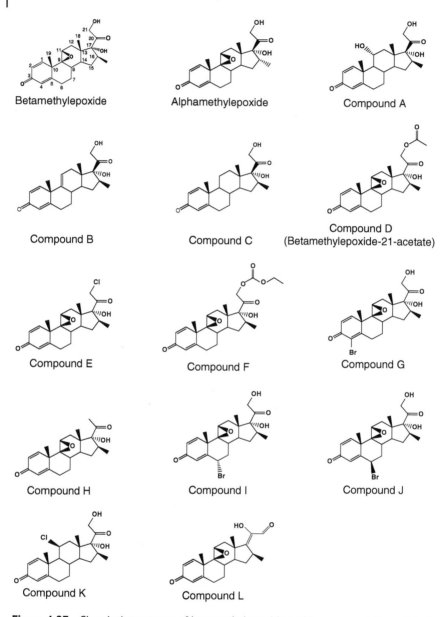

Figure 4.27 Chemical structures of betamethylepoxide and its representative related compounds [26]. *Source:* Reproduced with permission of Elsevier.

synthetic steps, or they can undergo similar reactions to form new impurities in the final products. The time and effort spent on the identification of the impurities and their sources, or the toxicity study of the new impurities can be saved if the type and amount of impurities in the raw materials are tightly controlled.

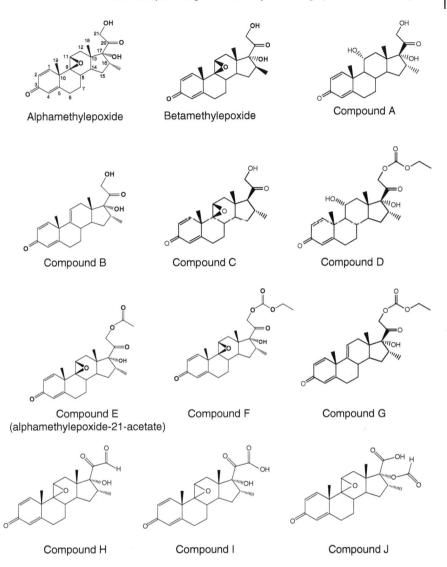

Figure 4.28 Chemical structures of alphamethylepoxide and its representative related compounds [28]. *Source:* Reproduced with permission of Oxford University Press.

As we already know that the separation of these structurally similar steroid compounds can only be achieved by finding suitable columns and combinations of mobile phase components. At this stage, there is really no secret but only hard work. As many as approximately 30 (for betamethylepoxide) and 50 (for alpha-methylepoxide) columns and a large number of combinations of the mobile phase compositions were tested. The end results are shown in Figures 4.29 and 4.30.

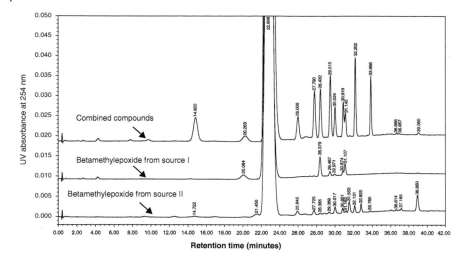

Compound names	Retention times in the top chromatogram	Retention times in the middle chromatogram	Retention times in the bottom chromatogram
Compound A	14.822	Below LOD	14.722
Alphamethylepoxide	20.263	20.084	Below LOD
Betamethylepoxide	22.8	22.8	22.8
Compound B	26.009	Below LOD	25.945
Compound C	27.790	Below LOD	27.725
Compound I	28.432	28.379	28.365
Compound H	29.515	29.467	29.369
Compound G	30.024	29.971	30.017
Compound D (betamethylepoxide-21-acetate)	30.919	30.874	30.851
Compound J	31.145	31.107	31.183
Compound L	Below LOD	Below LOD	31.553
Compound E	32.202	Below LOD	32.131
Compound K	Below LOD	Below LOD	32.820
Compound F	33.866	Below LOD	33.765
Dimmer	39.060	Below LOD	38.650

Figure 4.29 Typical chromatograms of (1) betamethylepoxide and its related compounds (top chromatogram, some of the related compounds were spiked at approximately 1% levels), (2) betamethylepoxide from source I (middle chromatogram), and (3) betamethylepoxide from source II (bottom chromatogram). The HPLC column is a 5 cm YMC Hydrosphere C_{18} (4.6 mm I.D.) and the mobile phase consists of water and acetonitrile : methanol (8 : 25, v/v) [26]. *Source:* Reproduced with permission of Elsevier.

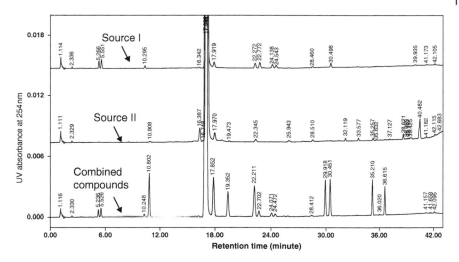

Compound names	Retention time in combined compounds (minutes)	Retention time in alphamethylepoxide from source II	Retention time in alphamethylepoxide from source I
Compound a	10.802	10.908	Not detected
Alphamethylepoxide	17.030	17.146	17.095
Betamethylepoxide	17.852	17.970	17.919
Compound b	19.352	19.473	Not detected
Compound c	22.211	22.345	22.272
Compound d	29.918	Not detected	Not detected
Compound E (alphamethylepoxide-21-acetate)	30.451	Not detected	30.498
Compound f	35.210	35.257	Not detected
Compound g	36.615	Not detected	Not detected

Figure 4.30 Typical chromatograms of (1) alphamethylepoxide from source I (top chromatogram), (2) alphamethylepoxide from source II (middle chromatogram), and (3) alphamethylepoxide from source I that was spiked with approximately 0.5% of the related compounds (bottom chromatogram). The HPLC column is a 15 cm ACE C_{18} (4.6 mm I.D.) and the mobile phase consists of 10 mM sodium sulfate, 0.05% (v/v) phosphoric acid, and acetonitrile [28]. *Source:* Reproduced with permission of Oxford University Press.

Similarly, to develop a method for the analysis of betamethasone and its related compounds (Figure 4.31), more than 10 columns and various combinations of acetonitrile, methanol, isopropanol, and THF were examined on each selected column. The hard work was paid off, as shown in Figure 4.32.

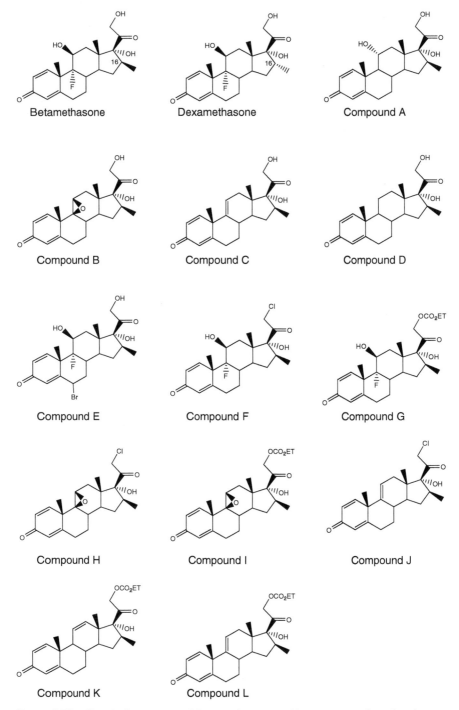

Figure 4.31 Chemical structures of Betamethasone and its representative related compounds [27]. *Source:* Reproduced with permission of Elsevier.

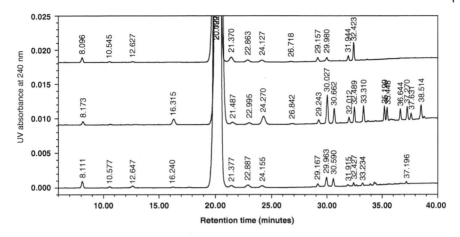

Compound names	Retention time in sample from source I (minutes)	Retention time in impurity profile sample (minutes)	Retention time in sample from source II (minutes)
Compound A	Not detected	16.315	16.240
Betamethasone	20.099	20.222	20.122
Dexamethasone	21.370	21.487	21.377
Compound B	24.120	24.270	24.145
Compound C	29.980	30.027	29.963
Compound D	Not detected	30.662	30.590
Compound E	32.423	32.489	32.427
Compound F	Not detected	33.310	33.234
Compound G	Not detected	35.199	Not detected
Compound H	Not detected	35.448	Not detected
Compound I	Not detected	36.644	Not detected
Compound J	Not detected	37.270	37.196
Compound K	Not detected	37.631	Not detected
Compound L	Not detected	38.514	Not detected

Figure 4.32 Typical chromatograms of (1) betamethasone from source I (top chromatogram), (2) betamethasone and its related compounds (impurity profile solution, middle chromatogram), and (3) betamethasone from source II (bottom chromatogram). This method employs an ACE Phenyl column (10 cm × 4.6 mm) and the mobile phase consists of (A) water : acetonitrile (90 : 10, v/v) and (B) acetonitrile : isopropanol (80 : 20, v/v) [27]. *Source:* Reproduced with permission of Elsevier.

References

1 Fischer, E. (1894). Einfluss der Configuration auf die Wirkung der Enzyme (Influence of the configuration on the effect of the enzymes). *Berichte der deutschen chemischen Gesellschaft* 27: 2985–2993.

2 Lehn, J.-M. (1987). Supramolecular Chemistry – Scope and Perspectives Molecules – Supermolecules – Molecular Devices, *Nobel Lecture*, December 8.

3 Gokel, G.W. (ed.) (1997). *Advances in Supramolecular Chemistry*, 4. JAI Press Ltd.

4 Bianchi, A., Bowman-James, K., and Garcia-Espana, E. (eds.) (1979). *Supramolecular Chemistry of Anions*. New York: Wiley.

5 Flieger, J. (2006). The effect of chaotropic mobile phase additives on the separation of selected alkaloids in reversed-phase high-performance liquid chromatography. *Journal of Chromatography A* 1113: 37–44.

6 Flieger, J. (2007). Effect of mobile phase composition on the retention of selected alkaloids in reversed-phase liquid chromatography with chaotropic salts. *Journal of Chromatography A* 1175: 207–221.

7 Méndez, A., Bosch, E., Rosés, M., and Neue, U.D. (2003). Comparison of the acidity of residual silanol groups in several liquid chromatography columns. *Journal of Chromatography A* 986: 33–44.

8 Manuel, J., Martínez, H., Méndez, A. et al. (2004). Characterization of the acidity of residual silanol groups in microparticulate and monolithic reversed-phase columns. *Journal of Chromatography A* 1060: 135–145.

9 Israelachvilo, J.N. (2011). *Intermolecular and Surface Forces*, 3e. Elsevier.

10 Luecke, H. and Quiocho, F.A. (1990). High specificity of a phosphate transport protein determined by hydrogen bonds. *Nature* 347: 402–406.

11 Jorgensen, W.L. and Pranata, J. (1990). Importance of secondary interactions in triply hydrogen bonded complexes: guanine-cytosine vs uracil-2,6-diaminopyridine. *Journal of the American Chemical Society* 112: 2008–2010.

12 Jorgensen, W.L. and Severance, D.L. (1991). Chemical chameleons: hydrogen bonding with imides and lactams in chloroform. *Journal of the American Chemical Society* 113: 209–216.

13 Bühlmann, P., Amemiya, S., Nishizawa, S. et al. (1998). Hydrogen-bonding ionophores for inorganic anions and nucleotides and their application in chemical sensors. *Journal of Inclusion Phenomena and Molecular Recognition in Chemistry* 32: 151–163.

14 Nishizaw, S., Bühlmann, P., Iwao, M., and Umezaw, Y. (1995). Anion recognition by urea and thiourea groups: remarkably simple neutral receptors for dihydrogenphosphate. *Tetrahedron Letters* 36: 6483–6486.

15 Bühlmann, P., Nishizawa, S., Xiao, K.P., and Umezawa, Y. (1997). Strong hydrogen bond-mediated complexation of $H_2PO_4^-$ by neutral bis-thiourea hosts. *Tetrahedron* 53: 1647–1654.

16 Xiao, K.P., Bühlmann, P., Nishizawa, S. et al. (1997). A chloride ion-selective solvent polymeric membrane electrode based on a hydrogen bond forming ionophore. *Analytical Chemistry* 69: 1038–1044.

17 McGaughey, G.B., Gagné, M., and Rappé, A.K. (1998). π-Stacking interactions. *The Journal of Biological Chemistry* 273: 15458–15463.

18 Lesellier, E., West, C., and Tchapla, A. (2006). Classification of special octadecyl-bonded phases by the carotenoid test. *Journal of Chromatography A* 1111: 62–70.

19 Rafferty, J.L., Siepmann, J., and Schure, M.R. (2009). The effects of chain length, embedded polar groups, pressure, and pore shape on structure and retention in reversed-phase liquid chromatography: molecular-level insights from Monte Carlo simulations. *Journal of Chromatography A* 1216: 2320–2331.

20 McCalle, D.V. (1998). Influence of sample mass on the performance of reversed-phase columns in the analysis of strongly basic compounds by high-performance liquid chromatography. *Journal of Chromatography A* 793: 31–46.

21 Dai, J., Carr, P.W., and McCalle, D.V. (2009). A new approach to the determination of column overload characteristics in reversed-phase liquid chromatography. *Journal of Chromatography A* 1216: 2474–2482.

22 Dolan, J.W. (2015). Overload in liquid chromatography. *LCGC North America* 33 (8): 528–533.

23 Liu, F., Xiao, K.P., and Rustum, A.M. (2009). Determination of individual homologues and total content of benzalkonium chloride by reversed-phase high-performance liquid chromatography using a short butyl column. *Journal of AOAC International* 92 (6): 1644–1651.

24 Xiao, K.P., Xiong, Y., Liu, F., and Rustum, A.M. (2007). Efficient method development strategy for challenging separation of pharmaceutical molecules using advanced chromatographic technologies. *Journal of Chromatography A* 1163 (1–2): 145–156.

25 Xiao, K.P., Xiong, Y., and Rustum, A.M. (2008). Quantitation of trace betamethasone or dexamethasone in dexamethasone or betamethasone active pharmaceutical ingredients by reversed-phase high-performance liquid chromatography. *Journal of Chromatographic Science* 46 (1): 15–22.

26 Xiao, K.P., Chien, D., Markovich, R., and Rustum, A.M. (2007). Development and validation of a stability-indicating reversed-phase high performance liquid chromatography method for assay of betamethylepoxide and estimation of its related compounds. *Journal of Chromatography A* 1157 (1–2): 207–216.

27 Xiong, Y., Xiao, K.P., and Rustum, A.M. (2009). Development and validation of a stability-indicating RP-HPLC method to separate low levels of dexamethasone and other related compounds from betamethasone. *Journal of Pharmaceutical and Biomedical Analysis* 49 (3): 646–654.

28 Xiao, K.P., Liu, F., Xiong, Y., and Rustum, A.M. (2009). Development and validation of a stability-indicating RP-HPLC method for assay of alphamethylepoxide and estimation of its related compounds. *Journal of Chromatographic Science* 47 (5): 378–386.

5

Degradation Chemistry and Product Development

The chemical stability of an active pharmaceutical ingredient (API) in a drug product is determined by its molecular structure, and by the environment in which the API molecule resides. Understanding the degradation chemistry facilitates (1) forming a proper formulation development strategy; (2) making right choices of raw materials; (3) selecting protective and yet economical packaging materials and configurations; (4) determining appropriate storage and transportation conditions; (5) defining proper in-use requirements; (6) coming up with reasonable prediction of potential shelf life of the formula; and last but not least, and (7) developing fit-for-purpose stability-indicating analytical methods.

Conducting stability studies is a regulatory requirement for any drug products that need registration [1]. In the United States, the FDA over-the-counter monograph system allows drug manufacturers to market new products without seeking FDA prior approval, provided that the API molecules belong to the group of molecules that are Generally Recognized As Safe (GRAS). Manufacturers of over-the-counter (OTC) monograph products typically consider their products are good as long as the products meet the specification acceptance criteria defined in corresponding USP drug product monographs. Occasionally, a product may contain multiple "GRAS" API molecules. The product is still regarded as an FDA monograph product. However, there may be no corresponding USP drug product monograph available for the API combinations intended for that product. Without a USP public standard or testing procedures, the manufacturers of that combination product are left alone or are free to use whatever methods developed in-house to test and release the product, and yet do not need the FDA's approval. Furthermore, even if there is a USP monograph available, sometimes the method is not state-of-the-art and is not able to monitor degradation products that are specific to those newly developed formulas. Since a drug manufacturer could unknowingly produce drug products that inadvertently contain impurities at

Analytical Scientists in Pharmaceutical Product Development: Task Management and Practical Knowledge, First Edition. Kangping Xiao.
© 2021 John Wiley & Sons, Inc. Published 2021 by John Wiley & Sons, Inc.

toxic levels, meeting the requirements for impurities listed in USP monographs is not necessarily adequate to ensure the safety of the consumers. Indeed, the quality assurance of OTC monograph drug products should not be treated any differently from that required for drugs marketed under an approved application.

During the degradation, lots of structural changes may happen to the API molecules. Some structural changes are easy to happen, which is referred to as primary degradation; some are secondary, i.e. degradation of degradation products and thus are less of a concern due to lower possibility or lower level of formation of such secondary degradation products. Some degradation products can also be the same molecules that formed (or is used) during the API synthesis and are often referred to as process impurities. Organic chemistry is the basis for the understanding of degradation chemistry, and this could potentially create some challenges to analytical scientists, who may not be very keen on the functional group reactivity and reaction mechanisms. This chapter hopes to describe some basic degradation chemistry, provide some example approaches to conducting forced degradation study, and to present known degradation pathways for some representative APIs in OTC products. For further studies, readers can refer to literature and books written by subject-matter experts [2–10].

Degradation of an API molecule is usually induced by one or more of the following conditions: (1) heat (dry or wet); (2) water or moisture (hydrolysis at various pHs including extreme acidic, basic and neutral conditions); (3) existence of oxidative species (oxidation); and (4) light (photolysis). To understand the intrinsic chemical stability of the API molecule, analytical scientists subject the drug substance to those conditions for a certain period of time to achieve a target degradation level of around 10–15% loss in API potency. Some literature or health authority guidelines recommend or request up to 30% loss in API potency, which may be too excessive. Among the possible degradation pathways, hydrolysis and oxidation are the most common culprits for drug instability.

Before moving on to the degradation chemistry, let us not forget that conducting the forced degradation is also a must for developing an analytical method and validating the stability-indicating power of testing procedures. The analytical method development should start with API forced degradation, followed by examining the stability-indicating power on formulation prototypes (preferably with placebo stressed in parallel), and aged samples, if available. Furthermore, forced degradation helps to set up appropriate release and shelf-life specifications for drug substances and drug products.

5.1 Hydrolysis

Hydrolysis is the most common pathway of pharmaceutical drug product degradation [11]. Especially, API molecules containing carboxylate ester, amide, or carbamate groups are prone to hydrolysis. For example, acetylsalicylic acid (aspirin) is readily hydrolyzed to form salicylic acid (Figure 5.1). The reaction is so fast that once the acetylsalicylic acid molecule is in contact with water, the hydrolysis happens instantaneously. The rate of hydrolysis is dependent on the pH. Both acid and base can catalyze the reaction, and neutral pH facilitates the formation of mixed products (Figure 5.1). Although aspirin is a more than a 100-year old molecule, the mechanism of hydrolysis of this molecule had many debates [12]. Interested readers can self-explore further.

Acetylsalicylic acid itself is not hygroscopic. The water in an acetylsalicylic acid containing drug product either comes from free water that exists during the manufacturing or from the moisture that is possessed or absorbed by the excipients. Again, always keep in mind that the behavior of a molecule is not only determined by its intrinsic chemistry but also is greatly influenced by the environment inside the dosage form where the API resides. The hydrolysis of acetylsalicylic acid under base conditions is much faster and much more complete than the hydrolysis under acidic conditions. The slowest reaction rate is at pH 2.5. L. J. Edwards et al. presented the effect of pH and temperature on the rate constant of

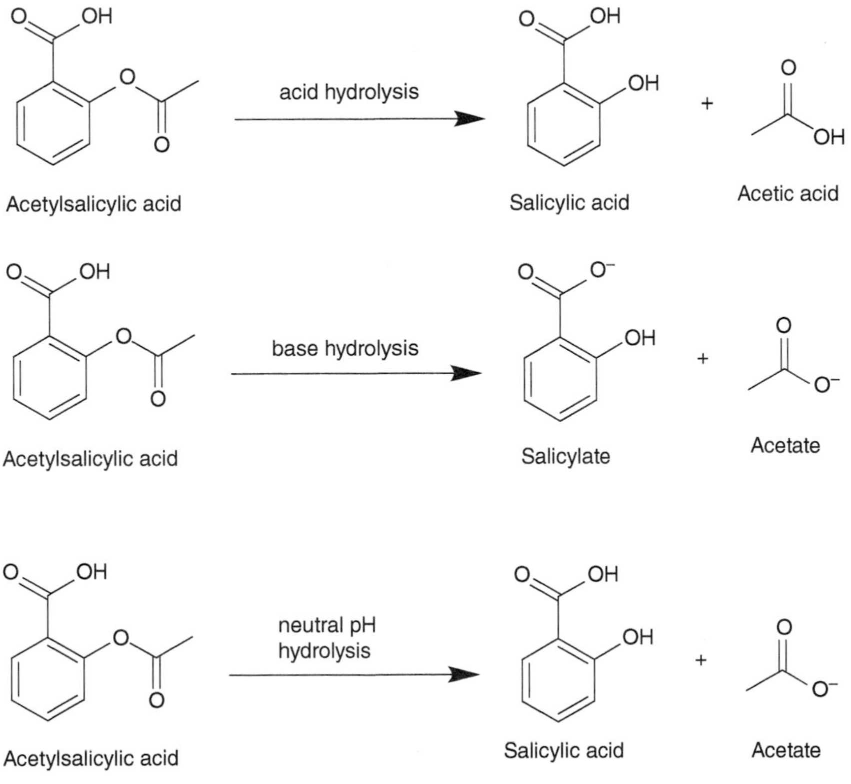

| Acetylsalicylic acid | acid hydrolysis | Salicylic acid | + | Acetic acid |

| Acetylsalicylic acid | base hydrolysis | Salicylate | + | Acetate |

| Acetylsalicylic acid | neutral pH hydrolysis | Salicylic acid | + | Acetate |

Figure 5.1 Hydrolysis of acetylsalicylic acid.

aspirin hydrolysis [13]. At 25°C, the reaction rate constant is about 10 times larger at pH < 1 than at pH 2.5, whereas the hydrolysis rate constant is more than a hundred times larger at pH > 11 than at pH 2.5. Between pH 4 and 8, the hydrolysis reaction of aspirin has a stable (unchanged) rate constant, which is slightly over one and a half times the rate constant at pH 2.5. The rate constant starts to curve up at pH 9 and takes off at pH > 11. The temperature effect is significant as well. At pH 2.5, the rate constant at 45°C is about two times as that of at 25°C. The difference between the rate constants at 45 and 25°C becomes about eight times at pH 5 [13]. Acetic acid forms during the hydrolysis (Figure 5.1), which is responsible for the vinegar smell of aspirin products when they are not appropriately handled during use or when the products are expired.

Another hydrolysis example is the amide hydrolysis of acetaminophen (the name of acetaminophen is commonly used in the United States, and the name of paracetamol is more common in Europe). Acetaminophen is widely prescribed (by doctors or by self-prescription) antipyretic and analgesic all over the world for treating the symptoms of different painful processes. It is also used as an intermediate for pharmaceuticals such as a precursor in penicillin [14]. The product of acetaminophen hydrolysis is 4-aminophenol (or para-aminophenol, PAP) [15] (Figure 5.2), whose safety limit has been established at 0.15% (w/w) relative to acetaminophen based on a thorough toxicological evaluation [16]. Although the hydrolysis of acetaminophen also produces acetic acid, the degree of hydrolysis of amide is much less than that of the ester group on acetylsalicylic acid, and therefore, the vinegar smell is usually not sensible. In the case of acetaminophen, the pH-rate profile for hydrolysis shows a V-shape curve with the lowest point at pH approximately 6 [15]. However, completely different from what happens to aspirin when being dissolved in water, acetaminophen can have a shelf-life in an aqueous solution that is estimated to be over three years at pH 6 and 25°C [15].

The third example is the carbamate group, which can be viewed as a mix of ester and amine groups. Loratadine, a nonsedative antihistamine compound that is

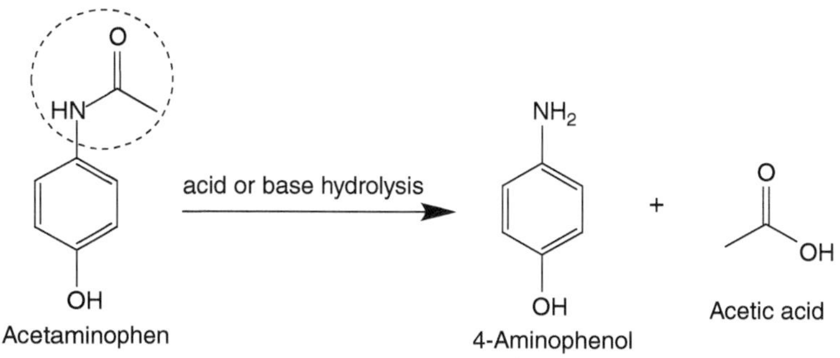

Figure 5.2 Hydrolysis of acetaminophen.

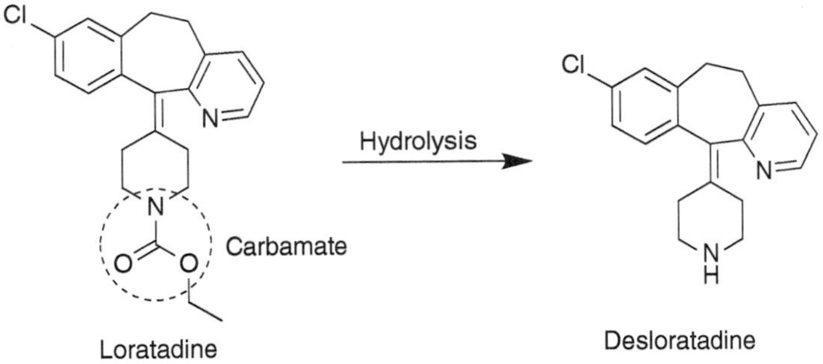

Figure 5.3 Hydrolysis of loratadine.

widely found in allergy relief drug products to help relieve allergy symptoms, such as sneezing, runny nose, and itchy, watery eyes, possesses a carbamate group (Figure 5.3) that can be hydrolyzed to form desloratadine, which itself is one of the APIs found in allergy relief products. Besides the formation of desloratadine, the other side products from the loratadine carbamate hydrolysis are ethanol (CH_3CH_2OH) and carbon dioxide (CO_2) [17].

5.2 Oxidation

It is reported that freshwater in an open vial contains approximately 6 mL/L of O_2 at 25°C, and about twice as much at 0°C [2]. Once more, as we have emphasized on many occasions in this book that the environment affects the stability of a molecule, the rate of oxidation is environmental pH- and temperature-dependent and can be catalyzed by metals and/or light. Different from inorganic oxidation–reduction reactions, where the electron gain and loss are clearly defined, the oxidation in organic reaction sometimes is not clearly accompanied by electron transfer. It is more commonly judged by the change in the oxidation state of the carbon atom. Another way of determining whether or not a reaction is oxidation is to see whether there is an oxygen atom being added to the molecule or if there is a hydrogen atom being abstracted from the molecule. An example of that is the oxidation of $-CH_2$ to C=O (ketone), with one O atom being added and two H atoms being removed. As a quick rule of thumb, the more stable radical forms as the result of losing an H atom, the easier the H atom can be abstracted from the molecule.

Oxidation has much more complicated mechanisms than other degradation pathways. Commonly observed oxidation reactions include (1) singlet oxygen oxidation; (2) autoxidation; (3) peroxide mediated oxidation; and (4) transition metal-induced oxidation.

There have been many top-level scientists on the journey of the discoveries of oxygen and its properties. They are Joseph Priestley, Amedeo Avogadro, Michael Faraday, and Robert S. Mulliken, only to name a few. Note that molecular oxygen, O_2, is not a powerful oxidation agent by itself. As shown in the molecular orbital structure in Figure 5.4, the anti-bond π orbitals contain one electron at each orbital that spins at the same orientation (so-called triplet oxygen) [18]. The molecular oxygen is thus considered as a diradical. However, since most of the organic molecules have paired electrons formed by overlapping of their valence atomic orbitals, it is less energy favorable for the triplet oxygen to directly react with those molecules (spin forbidden) [19]. When the triplet oxygen is stimulated by energy, such as by UV or light, the unpaired electrons in the triplet oxygen, shown at the left side of Figure 5.4, get the energy and become paired, leave one anti-bond orbital empty. The triplet oxygen becomes a so-called singlet oxygen [20], shown on the right side of Figure 5.4.

This "open house" arrangement then makes the oxidizing power of the oxygen molecule, i.e. grabbing electrons from other molecules, orders of magnitude stronger. The singlet oxygen is known to directly attack the C=C double bonds to form (hydro)peroxides or ketones (or aldehydes) (Figure 5.5) [21].

The singlet oxygen oxidation products are different from the free radical-induced triplet oxygen oxidation, which is referred to as the autoxidation. As shown in Figure 5.6, the autoxidation occurs at a different carbon position, the allylic carbon, as compared to the direct attack on the C–C double bond in Figure 5.5.

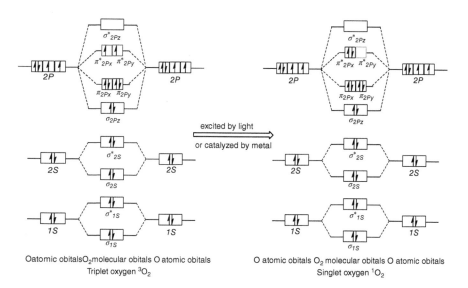

Figure 5.4 Molecular orbital diagram of triplet and singlet oxygen.

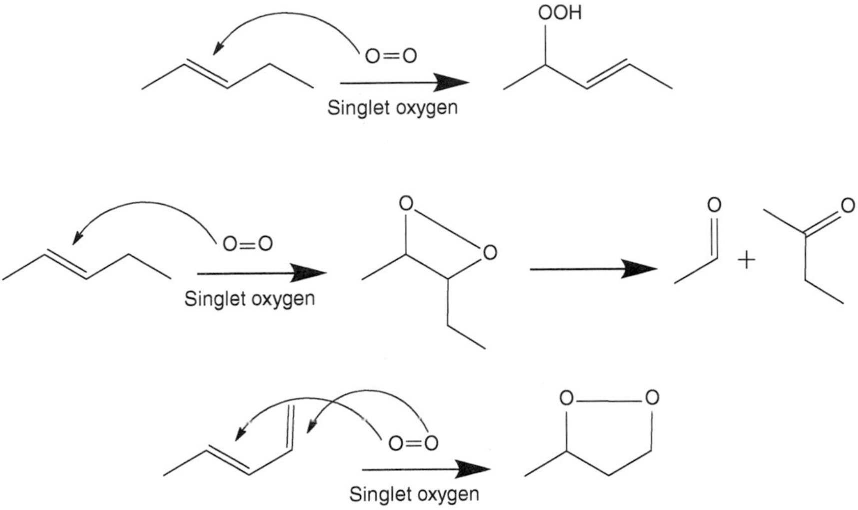

Figure 5.5 Singlet oxygen attacks directly on C–C double bond.

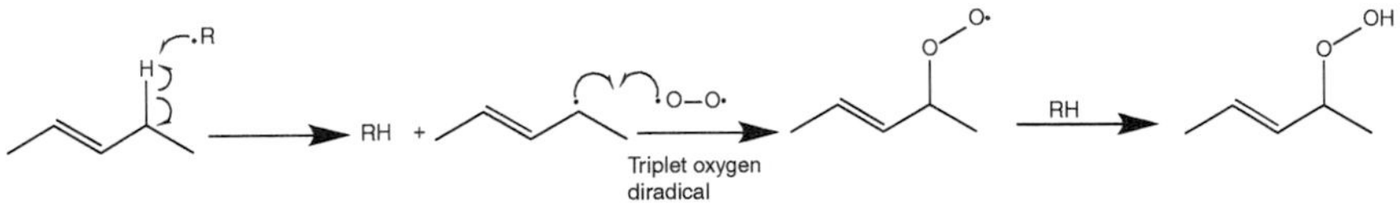

Figure 5.6 Triplet oxygen attacks the radical formed on an allylic carbon.

So now we know that autoxidation is the oxidation induced or prompted by the existence of molecular oxygen. The diradical structure makes the molecular oxygen reactive to other radicals. When there is a trace amount of organic peroxides (ROOR) in some of the ingredients of a drug product, with the help of heat or light, peroxy (ROO·) and oxy (RO·) radicals can form from the decomposition of ROOR. The peroxy and oxy radicals serve as the initiators to abstract hydrogen atoms from the drug molecule. The thus formed drug molecule radicals collide with the molecular oxygen to form peroxy radicals on the drug molecules, which keep reacting with other drug molecules to propagate the radical reaction or to form hydroperoxides. The abstraction of H atoms usually occurs on an allylic carbon (Figure 5.6), the alpha position of a heteroatom (Figure 5.7), and benzylic carbon (Figure 5.8).

In the case of cetirizine molecule, the H on the carbon that is shared by the 4-chlorophenyl and phenyl, and connecting to the nitrogen atom of the left side of the piperazine ring is very easy to be removed and leave behind a stable radical which subsequently forms 4-chlorobenzophenone (Figure 5.9).

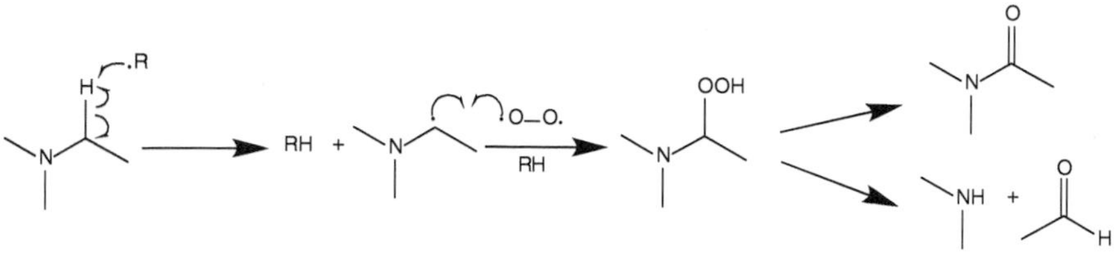

Figure 5.7 Scheme of autoxidation of the alpha-carbon near a heteroatom.

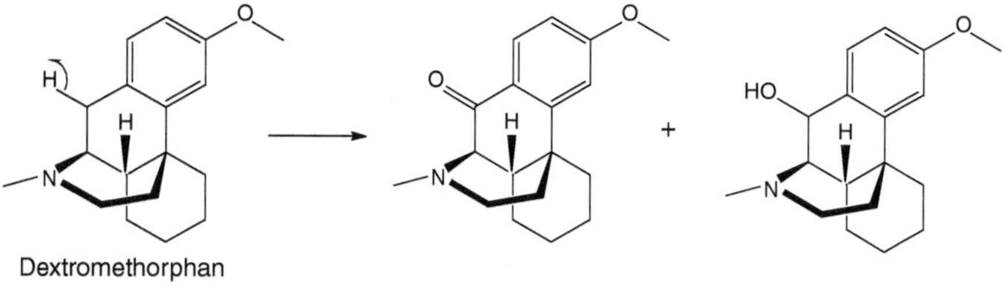

Dextromethorphan

Figure 5.8 Autoxidation of dextromethorphan, H atom abstraction on a benzylic carbon.

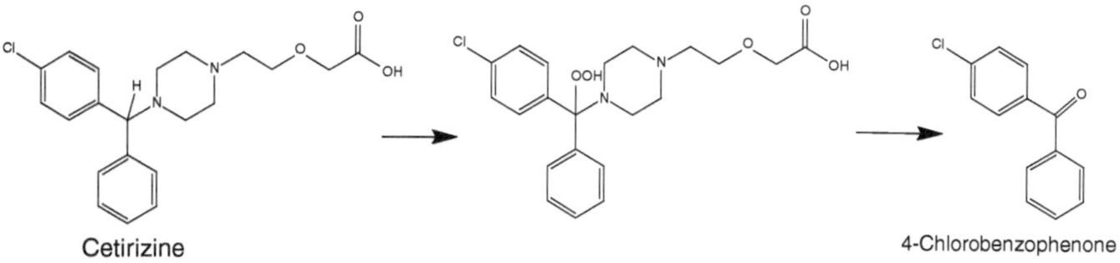

Cetirizine

4-Chlorobenzophenone

Figure 5.9 Most predominant autoxidation pathway of cetirizine.

Oxidation induced by peroxides accounts for another main oxidative degradation incidents of pharmaceutical products. Organic peroxides, including hydrogen peroxide H_2O_2, mainly come from excipients in the drug product. In addition, when molecular oxygen, either triplet oxygen or singlet oxygen, get one electron through single electron transfer from electron-rich functional groups, it becomes a superoxide radical anion O_2^-. This superoxide radical anion can form a hydroperoxy radical in water, HOO·, and can form a hydrogen peroxide HOOH, which can subsequently form a hydroxyl radical, HO·, in the presence of trace amount of transition metals.

One oxidation that is easy to predict is the N-oxide formation between tertiary amines and hydrogen peroxide. In this case, the hydrogen peroxide donates one oxygen (as O^+) to electron-rich functional groups such as amines. For example, the electron lone pair on the nitrogen atom attacks the H_2O_2 to form cetirizine N-oxide (Figure 5.10).

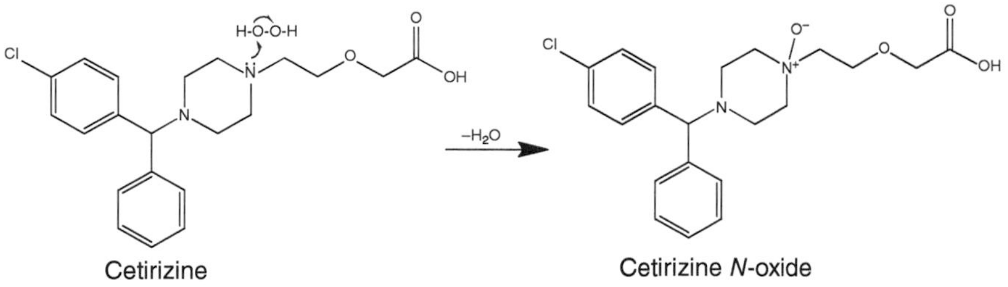

Figure 5.10 Cetirizine N-oxide formation.

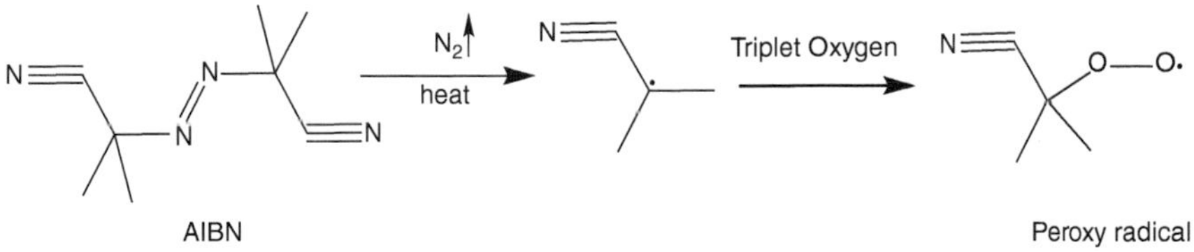

Figure 5.11 Schematic representation of radical initiation by AIBN.

Figure 5.12 Schematic representation of Fenton reaction.

$$Fe^{2+} + H_2O_2 \longrightarrow Fe^{3+} + HO\cdot + OH^-$$
$$Fe^{3+} + H_2O_2 \longrightarrow Fe^{2+} + HOO\cdot + H^+$$

It is a common practice during forced degradation studies, especially as part of the analytical method validation, to use hydrogen peroxide to assess the oxidation susceptibility of the drug API and drug product. The radicals formed from the decomposition of peroxides can induce the autoxidation as discussed previously. The predictive power, however, is not as prominent as using 2,2′-azobis[2-methyl-propanenitrilel] (AIBN) as a radical-chain initiator [22–24]. Attention should be paid, however, to the solvents used in the reaction [25] (Figure 5.11).

Oxidation catalyzed by metals is another common pathway in pharmaceutical product degradation. Metals like iron and copper are known to catalyze the oxidation of organic compounds through Fenton type reactions. When activated by metals such as Fe(II, III), hydrogen peroxide or organic peroxides break to form much reactive hydroxyl radicals (Figure 5.12).

The chemistry that is important to pharmaceutical product development is that it is not a straightforward relationship between the happening of Fenton type oxidation and the removal of those metal impurities by using chelating ligands such as ethylenediaminetetraacetic acid (EDTA), ascorbic acid, 1,2-dihydroxybenzene, etc. Adding chelating agents does not necessarily always help the stability of the product. The pro-oxidant and anti-oxidant effects of the ligands depend on the nature and coordination number. These factors, in turn, depend on the reaction conditions (pH, reagents concentration, etc.). The anti-oxidant ability of iron ligands has been associated with radical scavenger, radical deactivation, and inhibition of the initiation reaction by metal chelation. The pro-oxidant ability has been associated with the reduction of Fe^{3+} to Fe^{2+}, and the establishment of a redox cycle [26].

5.3 Reactions of Common Functional Groups

Although numerous chemical reactions are happening in nature, the good news is there are only a handful of common functional groups on OTC API molecules that are susceptible to degradation. They are (1) carboxylic acid (–COOH); (2) carboxylate ester (R_1COOR_2); (3) hydroxyl group (–OH); (4) unsaturated hydrocarbons (C=C); (5) amide group ($R_1CONR_2R_3$); and (6) amine group (–N; –NH, –NH$_2$). Common reactants that may exist as impurities in a drug product include aldehydes, formic acid, peroxides, etc.

5.3.1 Carboxylic Acid

Many pharmaceutical products, both APIs and excipients, contain –COOH, the carboxylic acid group. Understanding the reactions of carboxylic acid is useful for an analytical scientist to predict API-excipient compatibility and drug product stability. The most common reaction that the carboxylic acid gets involved is the esterification. An esterification reaction can happen in both acidic and basic environments when an –COOH group encounters a hydroxyl group (Figure 5.13).

Esterification can occur between API–API, API–excipient, and even API–solvent interactions. Figure 5.14 shows the esterification between cetirizine and a sugar alcohol, such as sorbitol, glycerol, mannitol, etc. [27, 28]. Esterification can

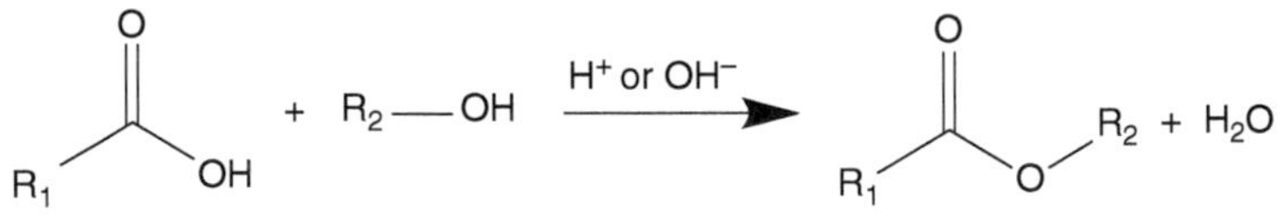

Figure 5.13 Schematic representation of esterification of carboxylic acid.

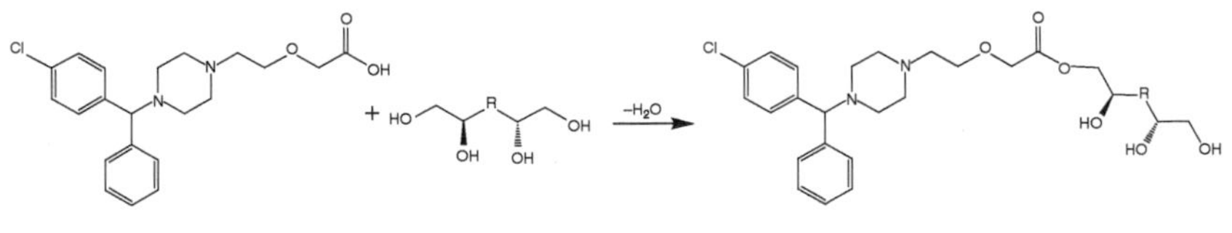

Figure 5.14 Esterification reaction between cetirizine molecule and sugar alcohol.

even occur when ibuprofen meets methanol (a mobile phase component) on the HPLC column (Figure 5.15) [29].

Figure 5.16 shows the esterification (and amide formation) between aspirin and phenylephrine. In this case, the reaction is called a *transesterification*. The esterification is not happening between the carboxylic acid group on aspirin and the hydroxyl group on phenylephrine. The esterification happens between the acetyl group of aspirin and the hydroxyl group of phenylephrine. The end result looks like the acetyl group and the hydroxyl group on the two molecules exchange their positions, and thus the reaction is called transesterification. Note that the acetyl group can also transfer itself to the amine moiety of phenylephrine and forms amide (the *N*-acetyl moiety) [30, 31]. As can be predicted, the transesterification can also happen between aspirin and acetaminophen.

Besides aspirin, methylparaben and propylparaben can react with sugars such as glucose, fructose, sucrose, lactose, maltose, cellobiose, and sugar alcohols such

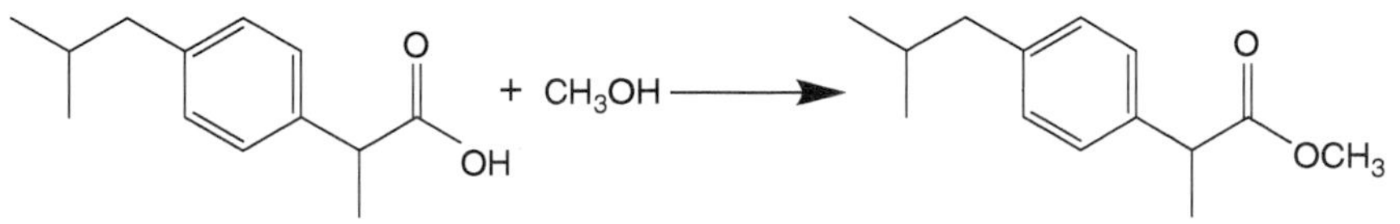

Figure 5.15 Esterification reaction between ibuprofen and methanol.

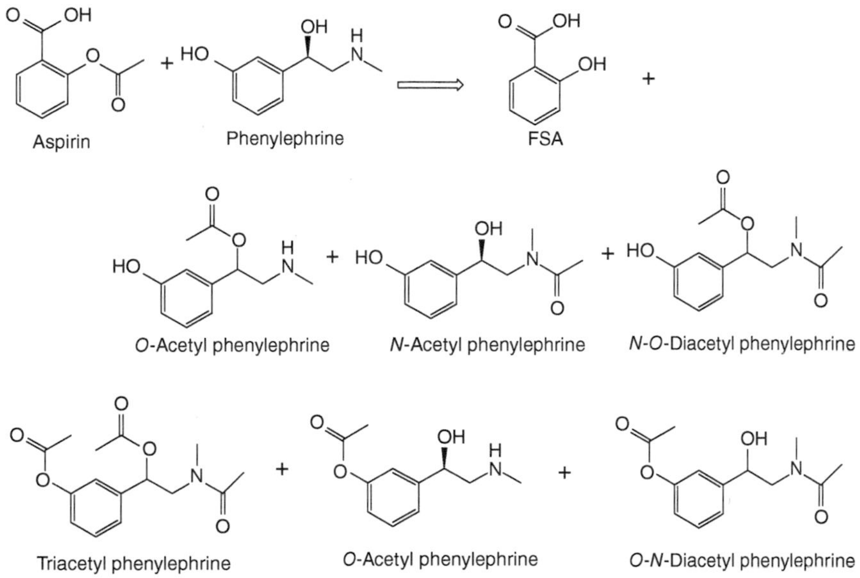

Figure 5.16 Transesterification reaction between aspirin and phenylephrine.

as lactitol, maltitol, sorbitol, mannitol, and polyols such as glycerol, through transesterification [32–34]. This transesterification is different from what is observed in the case of aspirin transesterification, that in this case, it is not the ester group that is transferred away from parabens. Apparently, it is a removal of the R–OH group, followed by a direct connection between the two molecules, as shown in Figure 5.17. Note that there are many –OH groups on the sugar and sugar alcohols. The reaction products thus have various isomers with different – OH group reacting with the parabens as long as the steric hindrance allows.

In some extreme cases, such as in a highly organic solvent environment, the carboxylic acid group can form anhydride. Figure 5.18 shows the formation of an aspirin anhydride. Reactions of an anhydride include the formation of acid, alcohol, and amide.

Decarboxylation is another degradation that the carboxylic acid containing APIs commonly undergo. A well-known decarboxylation is that naproxen decarboxylates to form various degradation products (Figure 5.19) [35, 36].

5.3.2 Hydroxyl Group

As shown in the above section, the hydroxyl group reacts with the carboxylic acid group to form esters. In addition, the hydroxyl group can also be oxidized. A primary hydroxyl group will be oxidized to form an aldehyde, and a secondary hydroxyl group will be oxidized to become ketone (Figure 5.20).

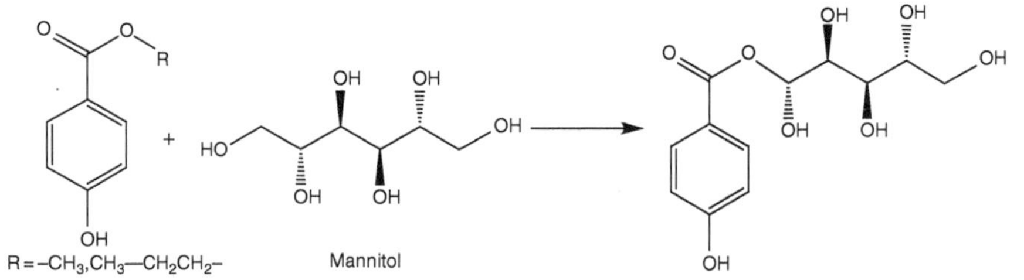

R = –CH₃, CH₃—CH₂CH₂–   Mannitol

R=-CH3,CH3—CH2CH2–

Figure 5.17 Schematic representation of methylparaben and propylparaben reaction with mannitol.

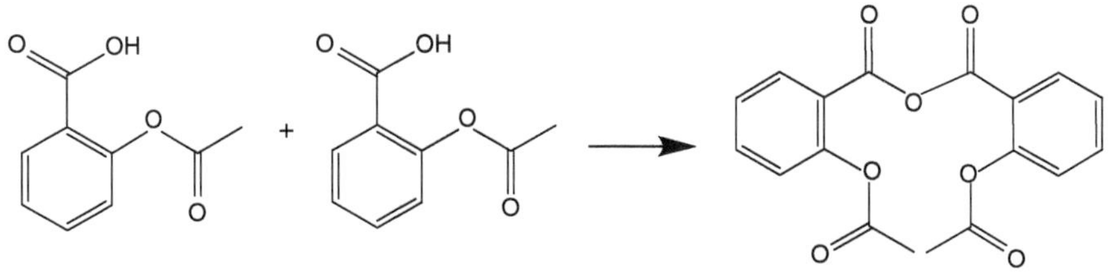

Figure 5.18 Anhydride formation between two aspirin molecules.

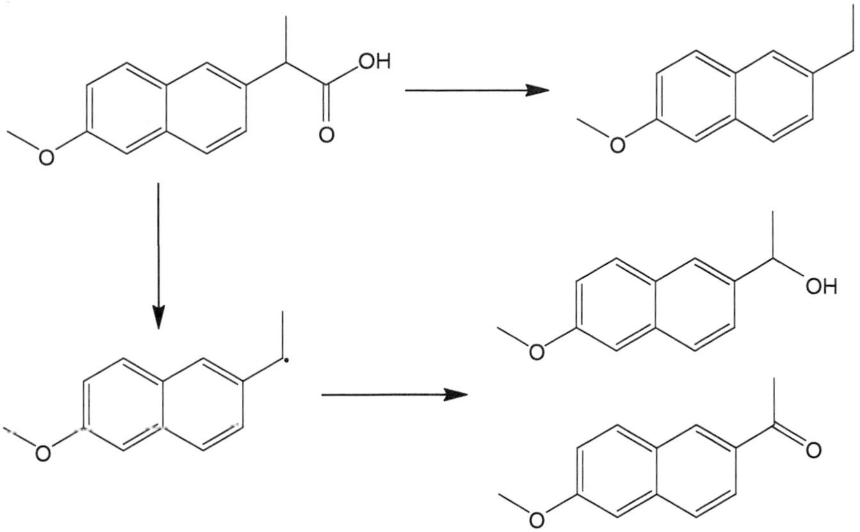

Figure 5.19 Decarboxylation of naproxen.

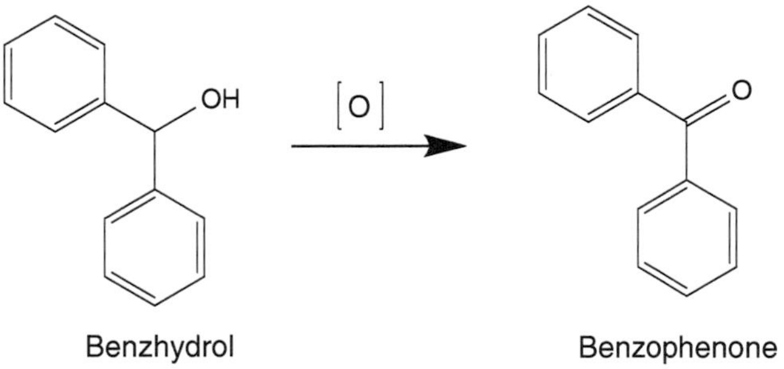

Benzhydrol Benzophenone

Figure 5.20 Oxidation of benzhydrol.

Another common hydroxyl reaction is the dehydration, i.e. removal of the –OH group. As an example, it has been reported that phenylephrine can lose its secondary hydroxyl group and forms the C=C bond (Figure 5.21) [37].

5.3.3 Carbon–Carbon Double Bond

A carbon–carbon double bond can carry many reactions. As mentioned previously, the C=C bond can be oxidized by singlet oxygen under photo stress (Figure 5.5), and autoxidation often happens at the allylic position since the allylic hydrogen can be easily removed (Figure 5.6). When being attacked by peroxyacids (or called peracids, RCO_3H), C=C can form epoxide (Figure 5.22). Nucleophilic

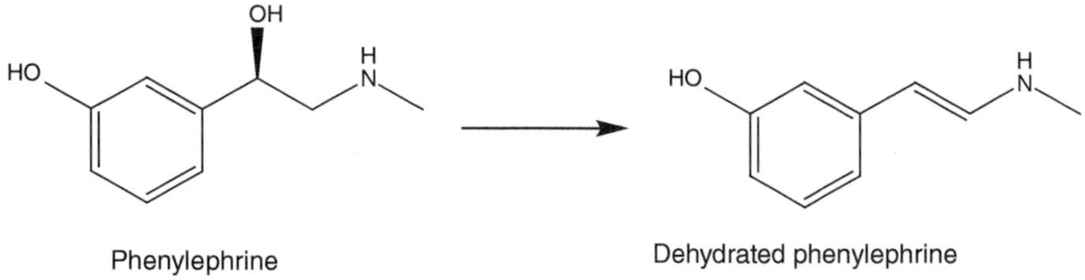

Phenylephrine

Dehydrated phenylephrine

Figure 5.21 Dehydration of phenylephrine.

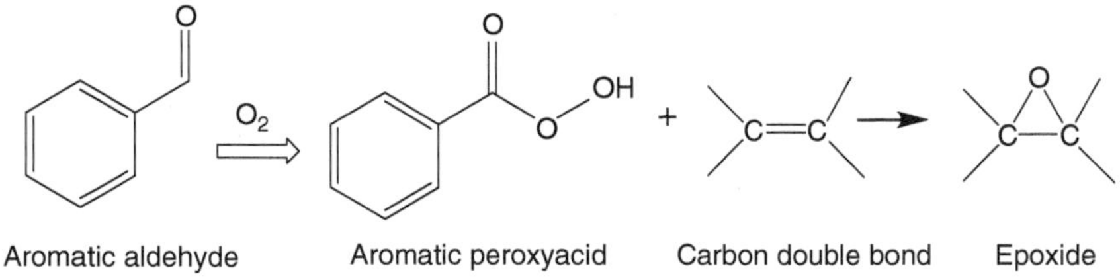

Aromatic aldehyde   Aromatic peroxyacid   Carbon double bond   Epoxide

Figure 5.22 Epoxide formation on a carbon–carbon double bond.

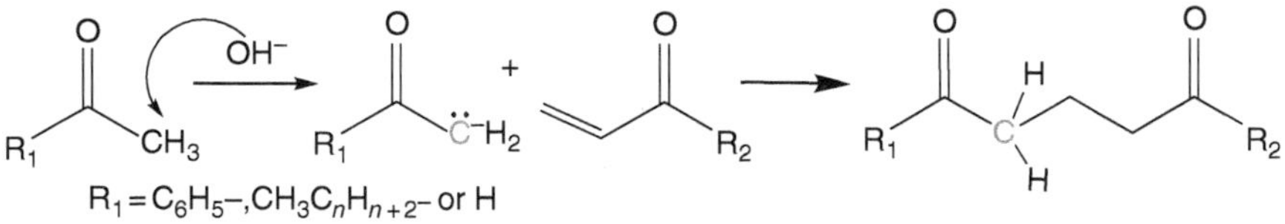

$R_1 = C_6H_5-, CH_3C_nH_{n+2}-$ or H

Figure 5.23 Michael addition between carbanion and α, β-unsaturated carbonyl compound.

moieties such as amines like to attack $R_1CH=CHR_2$ as well, and an amine adduct $R_1(R_3N)CH-CH_2R_2$ is formed as the result (Figure 5.26).

When the C=C double bond is part of α, β-unsaturated carbonyl compound (Figure 5.23), Michael Reaction can occur. This famous reaction is a nucleophilic addition to the C=C double bond by a nucleophile, such as a carbanion C^-, primary amine, secondary amine, and enamine.

5.3.4 Amine Reactions

Amine groups are common in molecular structures of various APIs. Understanding the amine reactions is critical for an analytical scientist to truly be able to help the formulation development to make stable products.

The key to understanding of amine chemistry lies in the acknowledgment of the electron lone pair on the nitrogen atom (N:). Various amines such as ammonia (N:H$_3$), tertiary amine ($R_1R_2R_3N$:), secondary amine (R_1R_2N:H), and primary amine

$(R_1N:H_2)$ can be protonated in solution to form NH_4^+, $R_1R_2R_3NH^+$, $R_1R_2NH_2^+$, and $R_1NH_3^+$, respectively. The electron lone pair on the nitrogen atom is responsible for nucleophilic attack from the amines to other electron "craving" functional groups, such as one of the oxygen in peroxides. The N-oxide formation shown in Figure 5.10 is a typical oxidation of APIs that have tertiary amines. Although not commonly observed in API degradation, primary amines can be oxidized to form hydroxylamine R_1NHOH, which can be further oxidized to form nitroso compound $R_1N=O$. Similarly, oxidation products for secondary amines are also hydroxylamine R_1R_2NOH, which can be further oxidized to form imine $R_1=NR_2$ or the nitroso compound as well. To minimize the amine reactivity, controlling the pH of the environment surrounding the amine molecules is critical. The environment should be acidic enough to protonate the nitrogen atom and thus "keep the electron lone pair busy and out of trouble."

The nucleophilic property of the electron lone pair on the nitrogen atom of amines makes the APIs very susceptible to degradation when there are flavors added to the formulas. Phenylephrine is notorious for such reactions [38]. Figure 5.24 shows the reaction between phenylephrine and formaldehyde. Other flavors may have other aldehydes such as benzaldehydes or acetaldehyde that can go through the same reactions and form corresponding degradation products.

One famous amine reaction is the Maillard reaction [39], which occurs between the primary or secondary amine group and the aldehyde moiety of reducing sugars such as lactose, fructose, and glucose (Figure 5.25). In some chewable products,

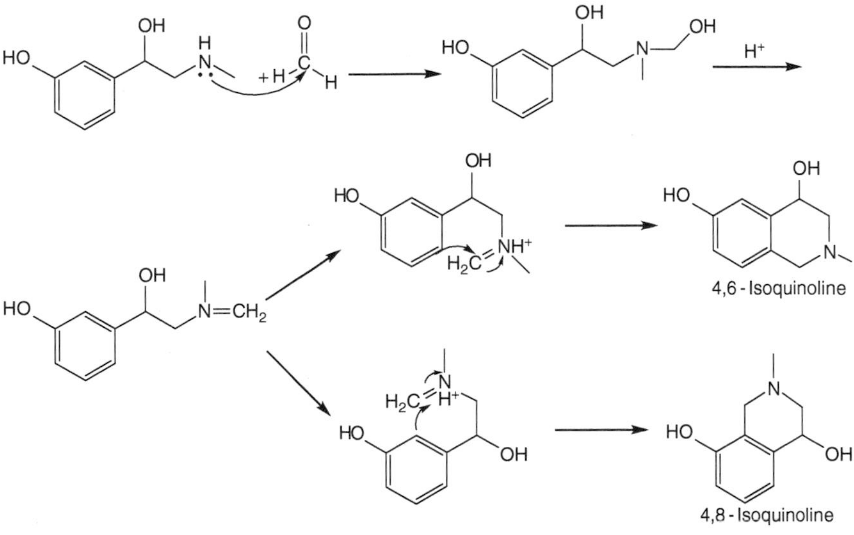

Figure 5.24 Phenylephrine reacts with formaldehyde and form 4,6- and 4,8-isomers.

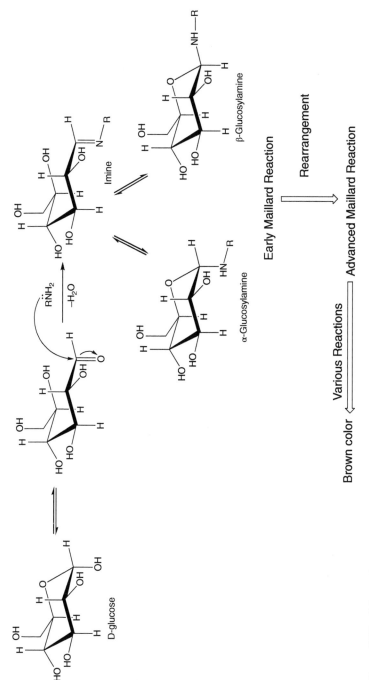

Figure 5.25 Schematic representation of Maillard reaction.

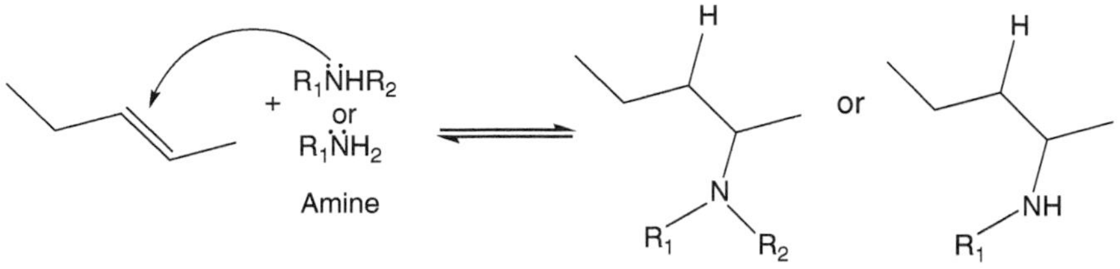

Figure 5.26 Amine reacts with carbon–carbon double bond.

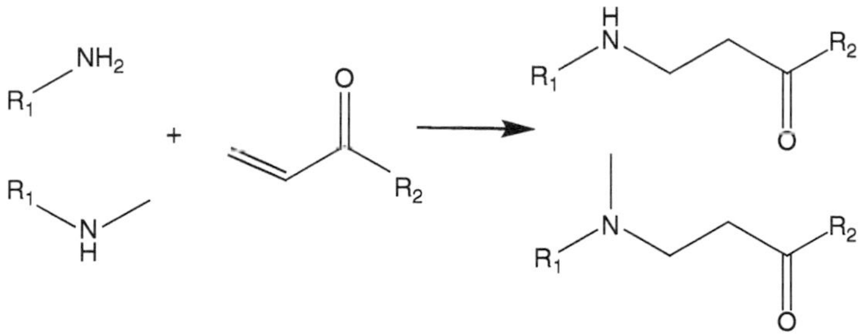

Figure 5.27 Michael addition between amine and α, β-unsaturated carbonyl compound.

Phenylephrine   Maleic acid

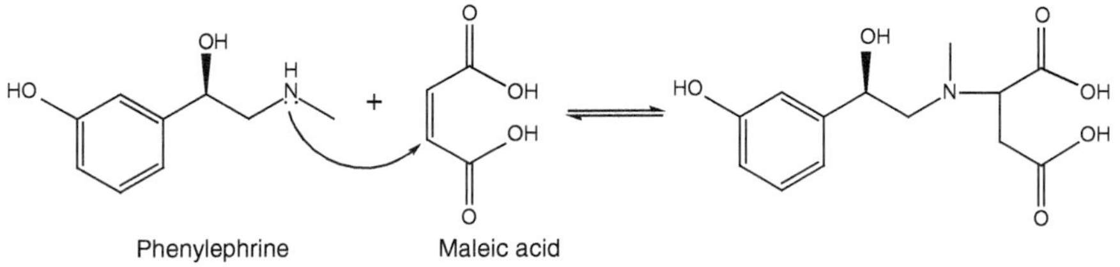

Figure 5.28 Michael addition between phenylephrine and maleic acid.

the Maillard reaction was observed between aspartame and a reducing sugar such as dextrose, which resulted in the loss of sweetness. Maillard reaction is a group of reactions that starts with amine meeting aldehyde. The browning color is attractive in the food industry and makes the cooking look appetizing, but it is not a piece of good news to the pharmaceutical drug product when browning color is observed by consumers or patients.

Amines can also launch a nucleophilic attack to carbon–carbon double bonds, which is called hydroamination (Figures 5.26 and 5.27).

As mentioned in the carbon double bond section, Michael reaction can happen between a primary or a secondary amine with the α, β-unsaturated carbonyl compound.

One good example of the Michael addition is the reaction between phenyle-phrine and maleic acid. Interestingly, this seemingly simple reaction, there are different opinions regarding the final product [40, 41]. The reaction scheme in Figure 5.28 is with which the author of this book agrees.

5.4 Summary of API Degradations

The aforementioned chemistry knowledge is very fundamental science that an analytical scientist should possess. Understanding those basics can help analytical scientists to contribute to product development from a chemistry point of view. An analytical scientist in pharmaceutical product development should be able to provide knowledge and not just a table of testing results. The scientist should be able to explain the happenings in the product and the stories behind the data, should be able to help the team to filter "bad players" from the excipient lists, and in return, also help him/herself to avoid testing a huge number of samples and work as a testing machine. Moreover, the ability to predicting potential degradation products helps the analytical scientist to develop stability-indicating chromatographic methods. The scientist may know what degradation products are potential watch-outs, so the chromatographic methods need to be ready for those peaks, should the degradation indeed occur.

Many pharmaceutical API molecules possess reactive functional groups or have structures that are susceptible to heat, light, moisture, oxidation, acidic, or basic stress. Those structures may be attacked by other molecules, radicals in the excipients, or light. The cetirizine molecule can serve as an example for us to review the basic knowledge described in this chapter (Figure 5.29). In Figure 5.29, the arrows are pointing to the places on the cetirizine molecule that chemical reactions can occur. The carboxylic acid group can react with hydroxyl (alcohol) groups (such as

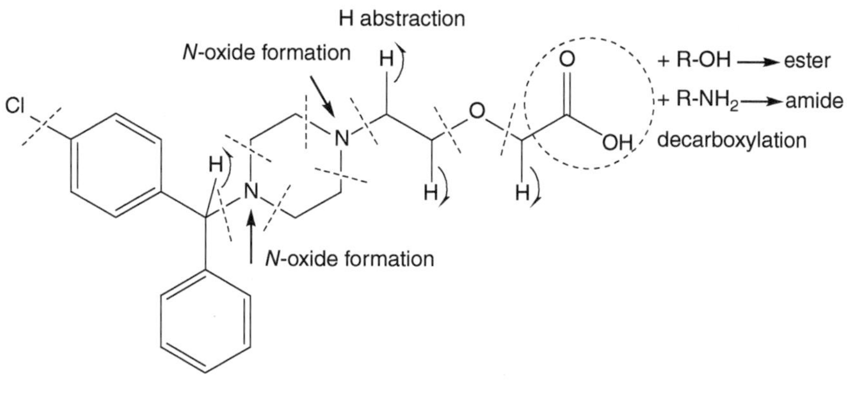

Figure 5.29 Degradation chemistry on cetirizine.

low-molecular-weight polyols or sugars) to form esters; can react with amines to form amides; and can decarboxylate by photolysis. The piperazinyl ring can react with peroxides to form N-oxides; can have ring opening to form ethylene diamine derivatives; and Millard reaction can occur once the secondary or primary amines are formed. The dibenzylic carbon is susceptible to autoxidation, peroxide mediated oxidation, and metal catalyzed oxidation to form 4-chlorobenzophenone. The carbon–chlorine, carbon–nitrogen, and carbon–oxygen cleavage can occur under light exposure, prolonged heating, metal catalyzed oxidation.

5.5 Stability Study and Forced Degradation

5.5.1 Guidelines on Long-Term Stability Study

Based on the ICH Q1A "Testing of New Drug Substances and Products," the purpose of pharmaceutical product long-term stability testing is to provide evidence on how the quality of a drug substance or drug product varies with time under the influence of a variety of environmental factors such as temperature, humidity, and light. The outcome of a long-term stability study is to establish a retest period for a drug substance or a shelf life span with recommended storage conditions for a drug product. The ICH guideline divides the weather conditions of the world into four climatic zones. Zone I climate is temperate, with a temperature and relative humidity (RH) at about 21°C/45% RH. Zone II climate is subtropical with possible high humidity, and a temperature and relative humidity at about 25°C/60% RH. Zone III climate is hot and dry, with a temperature and relative humidity at about 30°C/35% RH. Zone IV climate is further divided into Zone IV A, which is hot and humid at about 30°C/65% RH, and Zone IV B, which is hot and very humid at about 30°C/75% RH. The majority of the pharmaceutical industries conduct stability testing by storing the products in conditions according to those four climate zones. Items included in testing are chemical stability, physical performance, such as dissolution, disintegration, hardness, friability, etc., and microbial inertness of the products. Stability samples are pulled at predefined time points throughout the study. Other regulatory guidelines on stability studies include, but not limited to, ICH Q1B "Photo Stability Testing of New Drug Substances and Products," ICH Q2B "Validation of Analytical Procedures: Methodology," and WHO "Stability Testing of Active Pharmaceutical Ingredients and Finished Pharmaceutical Products." Many health authorities, including the FDA, adapt the practices described in those guidelines.

Typical conditions and terminologies for accelerated and long-term stability studies are listed in Table 5.1. The impurity identification threshold requested by ICH for API and drug product is listed in Table 5.2. In addition, the ICH M7 provides guidance on handling genotoxic degradation products.

Table 5.1 Typical long-term stability time and storage conditions.

Study	Storage condition	Minimum time period covered by data at submission
Long term[a]	$25 \pm 2°C/60\% \pm 5\%$ RH[b] or $30 \pm 2°C/65\% \pm 5\%$ RH or $30 \pm 2°C/75\% \pm 5\%$ RH	12 or 6 months
Intermediate[c]	$30 \pm 2°C/65\% \pm 5\%$ RH	6 months
Accelerated	$40 \pm 2°C/75\% \pm 5\%$ RH	6 months

[a] It is up to the applicant to decide whether long-term stability studies are performed at $25 \pm 2°C/60\% \pm 5\%$ RH or $30 \pm 2°C/65\% \pm 5\%$ RH or $30 \pm 2°C/75\% \pm 5\%$ RH.
[b] RH, relative humidity, is determined by the climatic condition under which the drug substance or drug product is intended to be stored.
[c] If $30 \pm 2°C/65\% \pm 5\%$ RH is the long-term condition, there is no intermediate condition.
Source: From ICH Guideline: STABILITY TESTING OF NEW DRUG SUBSTANCES AND PRODUCTS Q1A(R2).

Table 5.2 ICH impurity reporting, identification, and qualification thresholds for drug substance (DS) and for drug product (DP), based on total daily intake (TDI).

Maximum daily dose (g)	DS reporting threshold (%)	DP reporting threshold (%)
≤ 2	0.05	—
>2	0.03	—
≤ 1	—	0.1
>1	—	0.05
Maximum daily dose	**DS identification threshold**	**DP identification threshold**
<1 mg		1.0% or[a] 5 µg TDI
1–10 mg		0.5% or[a] 20 µg TDI
>10 mg–2 g	0.10% or[a] 1.0 mg TDI	0.2% or[a] 2 mg TDI
>2 g	0.05%	0.10%
Maximum daily dose	**DS qualification threshold**	**DP qualification threshold**
<10 mg	0.15% or[a] 1.0 mg TDI	1.0% or[a] 50 µg TDI
10–100 mg		0.5% or[a] 200 µg TDI
>100 mg–2 g		0.2% or[a] 3 mg TDI
>2 g	0.05%	0.15%

[a] Whichever is lower.
Source: From ICH Guideline: IMPURITIES IN NEW DRUG SUBSTANCES Q3A(R2), IMPURITIES IN NEW DRUG PRODUCTS Q3B(R2).

Based on ICH M7(R1) "Assessment and Control of DNA Reactive (Mutagenic) Impurities in Pharmaceuticals to Limit Potential Carcinogenic Risk" that was published in 2017, the Threshold of Toxicological Concern (TTC) concept was developed to define an acceptable intake for any unstudied chemical that poses a negligible risk of carcinogenicity or other toxic effects. For application of a TTC in the assessment of acceptable limits of mutagenic impurities in drug substances and drug products, a value of $1.5\,\mu g/day$ corresponding to a theoretical 10^{-5} excess lifetime risk of cancer, can be justified.

The guideline documents on stability studies published by the ICH or WHO provide only general guidance. The term "impurities" is used to refer to any compounds that are structurally related to the API, whether they are synthetic process impurities, degradation products of the API, or reaction products (adducts) between the API and other ingredients (excipients). The term can also include impurities in and degradation products of the excipients, residual solvents, and elemental impurities. One important but can be time-consuming task is the identification and quantitation of impurities observed during the stability studies. Establishing a prior knowledge space or a database within a product development R&D organization is helpful to facilitate formulation development, and ultimately to increase the product development speed-to-market. The knowledge space should contain the understanding of degradation pathways of various API molecules, insights of compatibilities of the API molecules with conventional excipients, and the awareness of kinetics of potential degradation reactions or API-excipient interactions. Such knowledge cannot be just obtained through long-term or accelerated stability studies. To obtain abundant information within the shortest possible time, purposefully forced degradation of API molecules is usually conducted. The conditions applied in the forced degradation study are much more severe than the long-term stability conditions including the accelerated conditions listed in Table 5.1.

5.5.2 Forced Degradation Study Considerations

Forced degradation or stress studies are essential for understanding the chemical stability of the API molecule(s) and the proposed formula(s). As stated previously, a lead analytical scientist should have a more than cursory understanding of organic chemistry. The scientist is expected to bring knowledge and not just data to the product development team. Oftentimes, analytical scientists in the OTC industry, however, find themselves conducting forced degradation studies on some molecules that have supposedly been studied before. The API molecules are well-known molecules, such as aspirin, naproxen sodium, ibuprofen, acetaminophen, etc. So, why conducting forced degradation again? Sometimes the answer from analytical scientists is that the previous forced degradation studies were conducted using a different or an out-of-date method. Or, previous studies were not conducted comprehensively that some conditions were too mild to see any significant degradation. Or, the studies were conducted some time ago, and the notebooks have been archived, and no report is available. As mentioned in Chapter 2, to share the knowledge and to achieve high efficiency in product development, platform/universal analytical procedures should be developed and are available for the whole analytical team to use. By the same logic, a carefully designed forced degradation protocol should be available for the team to follow. The experimental observations and obtained knowledge should be freely available among the analytical team members. To avoid forming silos in the team, and to avoid privatization of knowledge, the team should conduct the forced degradation in a consistent way; the experiments and results should be documented on a common platform (such as Electronic Laboratory Notebooks that can facilitate knowledge sharing based on meta-data search).

There is no one-size-fits-all approach in the pharmaceutical industry on how to conduct forced degradation studies. In addition to the variety of molecular structures of APIs, every company or even every scientist may have their own preferred practices and opinions on how should conduct the API forced degradation studies. There are many good literature works presenting various approaches to conduct forced degradation studies. This book will not go to any details to discuss those approaches and only list some of them as references [42–47]. The approaches described in Tables 5.3–5.6 are only examples for readers' information.

There are several points that the author would like to offer to the readers regarding the learning from the forced degradation study. Those points may be more valuable than the conditions provided in the tables. They are as follows:

1) The first thing to consider before starting the experimental work is the solubility of the API in the stress solutions. For example, stressing loratadine at high pHs without adding some organic co-solvent will generate faulty conclusions, such as loratadine is stable under basic conditions, while it is just because the API does not dissolve.

Table 5.3 Conditions and duration of the forced degradation study for thermal stress, with and without humidity.

Temperature	Duration	Single API	Dual APIs	Dual APIs
65°C	1, 3, 7 days	API in a closed glass container	—	API 1 mixed with API 2 followed by adding one to two drops of water to wet mix the APIs, in a closed and in an open glass container
65°C/75% RH	1, 3, 7 days	API in an open glass container	API 1 dry mixed with API 2 in an open glass container	—

Table 5.4 Conditions and duration of the forced degradation study for hydrolytic stress, at different pHs.

pH	Duration (day)	Media	Comments
1–1.2	1	0.1 N HCl	Neutralize with 0.1 N NaOH before making injection
5–6	1	20 mM Phosphate solution	—
9–10	1	0.1 N NaOH	Neutralize with 0.1 N HCl before making injection

Table 5.5 Conditions and duration of the forced degradation study for oxidative stress, with different oxidants.

Condition	Duration (day)
3% H_2O_2	1
10% solution of Tween 80 in water with 10 mM $FeCl_3$ at room temperature	1
10 mM $AIBN^a$ in 50 : 50 water : acetonitrile (v/v) at 40 °C	1

[a] 2,2-Azobisisobutyronitrile.

Table 5.6 Conditions and duration of the forced degradation study for light stress.

Condition	Duration
UV–vis	2× ICH^a

[a] ICH Q1B guideline requires exposure of samples to light providing an overall illumination of not less than 1.2 million lux hours and an integrated near ultraviolet energy of not less than 200 Wh/m². For stress study, a 2× ICH guideline requirement or above is recommended. Solid drug substances should be spread across the container to give a thickness of typically not more than 3 mm. Drug substances that are liquids should be exposed in chemically inert and transparent containers such as quartz glassware.

2) Degradation chemistry is different from the organic chemistry in that a yield of 0.5% in organic synthesis means practically the synthetic route does not work; while in degradation chemistry, a formation of 0.5% degradation product means a potential out-of-specification event.
3) Forced degradation studies conducted to support formulation development do not need to follow the conventional wisdom that the conditions should be able to cause a 10–15% degradation of the API molecule. The purpose of the forced degradation is, rather, to point out the potential degradation pathways of the API

molecule. If the API molecule is stable under some reasonable stress conditions, then the learning is sufficiently valuable for the formulation development, and no need to find harsher conditions to force the molecule to degrade more than 10%.

4) The practice of setting product specifications based on the forced degradation is not recommended. Sometimes those degradation product peaks are never observed in the real-life long-term stability studies of the product. To make things worse, when attempting to modify the method which contains some Specified Unidentified peaks that are only observed under some harsh stress conditions conducted by the original method developer, the effort of identifying those peaks (for example, in order to establish the method equivalency between the original method and the new method) can be enormous and unnecessary.

5) Mass balance is a very general topic when evaluating the analytical method specificity based on forced degradation studies. ICH defines mass balance as "the process of adding together the assay value and levels of degradation products to see how closely these add up to 100% of the initial value, with due consideration of the margin of analytical precision." Many health authorities demand mass balance from the submitted methods. However, from the basic chemistry point of view, mass balance should be replaced with molar balance. Many times, the molecular structures of some degradation products are unknown. The mass of the degradation products cannot be assessed and therefore not possible to calculate the mass balance. In addition, there are times the degradation products are adducts formed between the APIs and excipients. In those cases, the overall mass of the API, if adding all the API related compounds and the assay of API together, will then be more than 100%, which is not a balance, either.

Note that the ICH guideline considers that the minimum visible light exposure level represents approximately three months of continuous exposure to artificial visible light in the pharmacy, warehouse, or home with the protective container removed from the product. The UV light exposure roughly corresponds to one to two days inside close to a window with sunlight exposure [48].

5.6 Excipient Compatibility

5.6.1 General Remarks

A formula often contains many ingredients other than the API(s). Those non-API compounds are called excipients [49]. Their use in the formula is necessary for manufacturing the products to ensure the physical performances of the products, or to bring palatable organoleptic experiences to the consumers. According to the US National Formulary, excipients include chemicals that serve as acidifying/

alkalizing agent; aerosol propellant; antifoaming; antimicrobial preservatives; antioxidant; binder; buffering agent; bulking agent (freeze-drying); chelating/ sequestering agent; coating agent; coloring, flavor, perfume; diluent; disintegrant; emulsifying/solubilizing/wetting agent; glidant, anticaking agent; humectant; lubricant; ointment/suppository base; plasticizer; (co)solvent; stiffening agent; suspending/viscosity-increasing agent; sweetening agent; tonicity agent; and formula vehicle.

The excipients are defined as chemically inert ingredients. This notion is, however, usually misleading. Many excipients are chemicals that possess functional groups. Many of them have a trace amount of impurities that can cause instability of drug products. Drug–excipient interactions affect the API purity and efficacy, impact the drug product's organoleptic property, physical appearance, and cause performance changes such as slowdown in dissolution or increase in tablet hardness. It is a common observation that a lower drug–excipient ratio equals a higher risk in the occurrence of API degradation. It is not rare that if the API molecule is intrinsically unstable, it may experience even more degradation in the solid dosage forms than if the API is in its original (powder) form.

Excipients affect drug stability by one or combinations of the following means: (1) directly react with the API molecule; (2) provide reactive impurities that can either directly react with APIs or serve as catalysts in drug degradation; (3) bring moisture to the solid dosage form; and (4) create different microenvironment pHs in the dosage form. The knowledge of basic organic chemistry, degradation chemistry, and the mindset of thinking the overall physical/chemical environment in the dosage forms, together with knowledge of the chemistry of those potential reactive impurities in excipients, are needed to assess the risk of potential incompatibility between the API and the excipients.

5.6.2 Direct Reactions Between APIs and Excipients

Direct chemical reactions do occur between API molecules and some excipients. For example, APIs with carboxylic acid groups such as naproxen, aspirin, ibuprofen, cetirizine, etc., can react with the hydroxyl groups on sugars and sugar alcohols such as mannitol, sorbitol, sucrose, etc. It has been reported that cetirizine sorbitol and cetirizine glycerol esters were identified in oral liquid formulations, which were formed by the direct reaction between cetirizine and glycerol and sorbitol, respectively [27, 28]. Similarly, excipients having carboxylic acid groups, such as citric acid, can react with API molecules that have amine moieties. The Maillard reaction between the amine group of an API and the aldehyde group on reducing sugars is another example of those direct drug–excipient reactions [39]. The knowledge presented in previous sections can be found useful to predict or explain the degradation reaction happened between the API and various excipients, or to be more precise, between the API and the various functional groups on the excipients. Indeed, knowing the chemical structures and properties of the excipients is as important as knowing the API. Moreover, the chemistry knowledge of the excipients will help the analytical scientists to develop appropriate sample preparation procedures to minimize the amount of the excipients dissolved in the analytical sample solution, to minimize the retention of the excipients on the HPLC column, and to minimize the interference from the excipients (although sometimes the excipient peaks may not be obviously observed at the UV wavelength selected for monitoring API and its related compounds).

5.6.3 Impurities in Excipients

The term *excipient compatibility* can be misleading, however, since in most cases excipients do not directly form chemical bonds with API molecules but can directly cause degradation. It is the trace level of impurities left in the excipients during manufacturing or formed during the storage of the excipients that react with APIs or catalyze the degradation of the API molecules [47, 50–56]. Among those impurities are peroxides, aldehydes, acids, and metals. More specifically, aldehydes in flavors; aldehyde functional groups in reducing sugars; organic peroxides and hydrogen peroxides in polymeric excipients; nitrates; nitrites; heavy metals; and residual solvents [50, 51].

Hydroperoxides are common trace level impurities that exist in polyethylene glycol (PEG), povidone, crospovidone, hydroxypropyl cellulose, and polysorbate. Whether or not the peroxides exist in the form of ROOR or ROOH, they are highly reactive. They can react directly with API molecules or they can break down to even more reactive radicals, such as hydroxyl (HO·) and/or alkoxyl radicals (RO·), to induce severe oxidation reactions. The ROOR and the free radicals can also form other reactive oxygen species, such as superoxide anion ($O^{2-}\cdot$), hydrogen peroxide (H_2O_2), and organic hydroperoxides (ROOH). Some excipient, such as croscarmellose sodium, is reported to be able to act as a scavenger for peroxide. One additional thing to keep in mind is that free radical-induced oxidative drug degradation often shows an initial lag phase before the degradation becomes apparent.

Formic acid, formaldehyde, acetaldehyde, benzaldehyde, etc. can be present in flavors or PEG as trace impurities. Drugs with an amine group or hydroxyl group can react with formic acid or formate impurity to form amides or esters, respectively. Formaldehyde and formic acid can react with amine groups to form *N*-methyl or *N*-formyl adducts, and acetaldehyde can react with amine groups to form *N*-acetyl adducts. In addition, aldehyde impurities in excipients are commonly known to affect the disintegration of capsule gel formulations by chemical crosslinking of gelatin. Capsules usually show slower dissolution after storage if any aldehyde containing or contaminated excipients exist in the formula.

Not every sugar is a reducing sugar. For example, sucrose is a disaccharide and nonreducing sugar. However, such an "inert" material can degrade to generate glucose and fructose. The degradation is faster at pH 4 than it is at pH 6 or 7.

Since the impurities are at the trace levels, they could potentially be not well controlled during the manufacturing of excipients. On top of that, it is often not easy for the new product development team to obtain the exact chemical information about the potential trace amount of impurities, as the manufacturing processes of excipients are proprietary information. To make things more difficult, impurities, such as peroxides, can be generated during the storage of excipients or the finished product itself. Unfortunately, common malpractice in the new product development is that the excipient quality is thought to be adequate as long as the excipient has been released after being tested against the compendial requirements, such as USP/NF and Ph.Eur. Furthermore, for the proprietary excipients, such as flavors or colors, there is even less opportunity to scrutinize the chemical components, where aldehydes, peroxides, or metals are known to exist. Therefore, it is necessary to conduct excipient–API compatibility studies to minimize any future surprises in the drug product stability. Again, the knowledge of the basic reactions between those common functional groups with aldehydes, peroxides, metals, oxygen, etc., that are presented in previous sections, becomes handy.

The above chemical reactions or incompatibilities may not happen, however, if there is no water around or there is no suitable microenvironment to facilitate the degradation. For example, sodium carbonate or sodium bicarbonate in aspirin effervescent tablets can cause the hydrolysis of acetylsalicylic acid due to their ability to absorb water/moisture. Hydrolysis of Na_2CO_3 or $NaHCO_3$ significantly brings up the pH (>8) of the microenvironment surrounding the acetylsalicylic acid molecules, and thus de-esterification happens and salicylic acid forms. If mixing those excipients with aspirin under extremely dry conditions, the hydrolysis may not occur or occurs at a rate that practically does no harm to the product stability.

5.6.4 Solid-State Stability – Role of Water

At high temperatures under dry conditions, many API molecules are actually quite stable. For example, dry heating acetylsalicylic acid at 80°C for a few days may not result in any significant degradation of this molecule. While the story becomes very different when heating the compound at even moderate temperatures, such as 50°C, with some moisture around. Although purely solid-state degradation of drug substances is feasible, such instances are rare. Most drug degradation reactions in solid dosage forms involve moisture [57, 58]. From a chemistry point of view, degradation pathways an API molecule goes through in solutions are likely the same or highly similar to those observed in the (moisturized) solid-state. Having moisture in the excipients, introducing moisture during the manufacturing of the drug product, or absorbing moisture during the storage not only potentially degrade the products but also potentially slow down the dissolution or disintegration of solid dosage forms. From a physical-chemistry point of view, water/humidity/moisture-induced pharmaceutical product stability declining or performance deterioration is mainly caused by water stimulated physical effects such as plasticization rather than any direct reaction with water [59, 60]. Those physical changes include polymorphic transition, hydrate/solvate formation, dehydration or desolvation, crystallization of amorphous materials, vaporization, etc. For example, if an excipient, such as a flavor chemical, meets with water, it may lower its glass transition temperature (T_g) to be less than 40°C, which is the temperature of the accelerated stability condition. The excipient can thus change from a rigid glassy state into a movable liquid-like state. This physical change of the excipient at 40°C can subsequently hamper the release of the API from the product by retarding the disintegration or erosion of the product, and results in a slowdown of the product dissolution. On the other hand, dissolutions of the samples stored at lower temperatures such as 30 or 25°C do not show any slowdown because there is no such transition occurs at lower temperatures. Example excipients that can lower T_g of APIs are starch, lactose, and cellulose. In short, water is the most critical factor, not one of the factors that influence the chemical stability of API molecules and the performance of products.

Product development scientists need to know the difference among four concepts, the first one is the total water content, the second one is the absorbed water, the third is the hydrated water, and the fourth one is the free water. Loss on drying and Karl Fisher water content measurement are different in that the former one is about the water absorbed and the latter one is about the total water content including both bound (e.g. water of hydration) and absorbed water. It is the absorbed water that possesses the potential to impact drug product stability. However, one further step, which is the most critical step, has to be considered when assessing the impact of the absorbed water on product stability. It is the free water that is responsible for the degradation of moisture-sensitive materials. If the absorbed water is held firmly by the molecule that absorbs water, the water is then not available for the surrounding microenvironment. On the other hand, if the absorbed water can freely move around, thus given the name of free water, then those water molecules can change the surrounding microenvironment in the product. The amount of free water in a drug product, or an ingredient, is measured and expressed by water activity. Physical chemistry defines the water activity as the ratio of the partial vapor pressure in equilibrium with a substrate (e.g. a drug product or an ingredient) to the partial vapor pressure of pure water at the same temperature. Water activity is expressed on a scale of 0–1, where 1 is for pure water. The water activity has a more familiar term, that is, relative humidity, %RH. The absolute amount of free water available at the same relative humidity is different at different temperatures. The higher the temperature is, the more amount of free water is available at the same % RH. For example, the amount of free water available in the atmosphere with a 75% RH at 40°C is close to two times the amount of water at 30°C or nearly four times the amount of water at 25°C. With this clear differentiation between total water content, absorbed water, versus free water/ water activity in mind, a product development scientist should now understand the role of water is oftentimes to cause physical changes of the environment inside a dosage form. This behavior is related to the water activity of the drug product rather than the total water content in the product. Drug product water content will directly correlate with the degradation rate only if the water content correlates directly with the water activity. Therefore, somewhat counter-intuitive, since it is the free water that can cause product degradation, excipients that strongly absorb water can actually help the product stability by keeping water with them and do not allow the water to be freely movable. Examples of such excipients are starch, colloidal silica, and silica gel. On the other hand, water in excipients such as microcrystalline cellulose is highly reactive because it is weakly absorbed. It was found that aspirin hydrolyzed faster in the presence of microcrystalline cellulose (MCC) versus microfine cellulose (MFC) [61].

Another common means the excipients can impact the API stability is by altering the pH of the microenvironment in solid dosage forms. When water molecules exist, the excipients with ionizable functional groups can change the microenvironment pH due to the hydrolysis of themselves, a very similar behavior as if the excipients are dissolved in an aqueous solution. These pH-changing excipients include sodium carbonate, sodium bicarbonate, calcium carbonate, croscarmellose sodium, sodium starch glycolate, dicalcium phosphate, magnesium stearate, stearic acid, etc. The surface pH of one excipient is a good indicator of how much the excipient may impact the neighborhood pH in solid dosage forms. The surface pH can be measured simply by using a piece of pH paper to measure a wetted excipient or a slurry of that excipient [62]. It has been reported that the presence of magnesium stearate increases the pH of the microenvironment and accelerates the hydrolysis of acetylsalicylic acid [63]. For liquid dosage forms, the pH of the final solution should be controlled. The product development scientists should take into account the pH-changing ability of the dissolved excipients when designing the formulas.

An interesting thing to keep in mind is that in the solid dosage forms, even without water being there to serve as a solvent, hydrolysis of ester and amide can be facilitated by sugar and sugar alcohols such as dextrose, sucrose, sorbitol, or mannitol [8]. Some researchers argue that the reaction is attributed to the more nucleophilic alkoxy anion (RO^-) rather than the less nucleophilic unionized hydroxyl group (ROH). Nonetheless, as far as an analytical scientist is concerned, the analytical method may need to be ready for the potential formation of hydrolyzed degradation product of the API, and to provide this knowledge to formulation colleagues as a potential watch-out for API potency loss.

Finally, although water is definitely a critical component in drug degradation, keep in mind that, except for the direct hydrolysis, the main job that water molecules do is to help the API and excipient molecules move around. Even without water, molecular mobility can be increased by damaging the crystal lattice of the API. The use of abrasive excipients, such as colloidal silicon dioxide, or the application of strong shear or compression forces during granulation or compression process, respectively, can disturb the API crystallinity. Furthermore, in some extreme cases, the hydrated water can be lost from the excipients during grinding and the water facilitates drug degradation [64]. Since the acceptable specification limit of a degradation product is typically quite low in pharmaceutical dosages, a small region of crystallinity defect in the API molecule is sufficient to cause the drug product degradation to be out of specification.

5.6.5 Experimental Considerations – Formulation Relevancy

Analytical scientists in a product development team lead the project from a chemistry point of view. The advice, suggestions, and interpretation of the analytical data that the scientists provide to the team need to be thoughtful, constructive, and to move the project forward. Forced degradation and excipient compatibility studies are conducted at conditions much harsher than those applied during the accelerated stability evaluation defined by the ICH guidelines. As described before, the purpose of conducting forced degradation studies on API molecules is to understand the chemistry of the molecules, their intrinsic chemical stability, and in parallel, the analytical scientists are trying to build a stability-indicating analytical testing method [65–67]. After that stage, the purposes of conducting excipients compatibility studies are (1) to find out warning signs at early stages of formulation development, (2) for the analytical scientists to record any chromatographic peaks from the excipients, peaks from degradation of the excipients, peaks from the interaction between the API and excipients (adduct formation), and peaks from the API degradation found during the compatibility studies. Before moving on, it is necessary to clear one potential confusion between an excipient compatibility study and a so-called $N-1$ study intended for formula troubleshooting. If the task is to find out what went wrong in an existing marketed product and to see whether the formula can be saved by simply removing one or a few "bad player" excipients, the $N-1$ approach may have its merit. If the purpose is to develop a new formula, listing all the $N-1$ combinations among all the potential ingredients without knowing excipient compatibility beforehand, is a rather inefficient and ineffective approach.

As a side note, some forced degradation approaches, such as mixing the finished products with acid, base, or oxidants, however, not only offer little information regarding degradation kinetics but also potentially form irrelevant degradation products. The obtained information may not reflect real-life chemistry and may not sound relevant to the formulation scientists. Moreover, a formulation should include two aspects, one is the formula, and the other one is the process. The formula developed in a formulation laboratory without consideration of the future manufacturing process to make it scalable is not a true formula and thus is not a true formulation development. Therefore, before the process is being developed, conducting a full-course forced degradation on the early-stage formulation prototypes with invasive approaches such as soaking the products in acid or base is not only inefficient but also causes information overload.

There are many vastly different approaches for conducting excipient compatibility studies [68–71]. Just like the design of API forced degradation studies, there is no one-size-fits-all approach. Some even criticize the value of conducting such an excipient compatibility study [72]. Every formula is unique, and each API molecule has to be examined from its chemical structure, reactive functional groups, and known reactions with other excipients, as reported in the literature. In-house knowledge and experiences, potential manufacturing process, the intended use of the product, and the climate zones in which the product will be marketed are all part of the considerations. The results from the excipient compatibility study should be evaluated on a case-by-case basis, the information should be treated with a grain of salt rather than taking it as a yes or no decision, a black or white answer. As has been emphasized many times, the purpose of analytical development is to facilitate product development. Therefore, having some brief knowledge of the formulation/manufacturing processes will help analytical scientists to design suitable excipient compatibility approaches. Figure 5.30 illustrates a generic flowchart of a solid dosage form manufacturing process. In Figure 5.30, there are many steps involving dissolving, drying, mixing, blending, grinding, and compression. As shown in the flowchart, water addition into the formula by such as spraying, wet granulating, and coating during the manufacturing process is very common.

As we already know that during storage, the finished product can potentially absorb moisture and starts water-induced degradation. Based on the evaluation of many literature reports, the drug–excipient compatibility testing method adopted by Serajuddin et al. is the one that the author would recommend in this book [68]. The characteristic approach in the experimental design is to add 20% (v/w) water to the API-excipient mixture, followed by storing the mixtures at 50°C for one to three weeks. Of course, depending on the real cases, the time of stress can be shorter or longer, and the amount of water added can be more or less. Some literature work reports approach in which the moisture is incorporated in the system by exposing samples to high humidity, up to 100%, instead of adding water to the

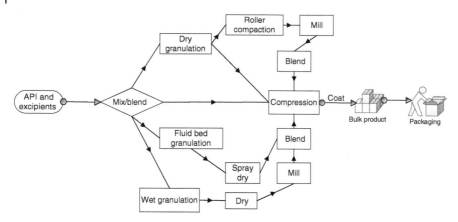

Figure 5.30 Schematic representation of pharmaceutical manufacturing processes for production of solid-dosage forms.

mixture. It was considered that the variability brought by the different hygroscopicities of drug substances and/or excipients might influence the compatibility study outcomes. The addition of a predetermined amount of water removes this unpredictability from the system. A detailed description of the procedures can be found in the original publication [68].

As a side note, sometimes to assess the mechanic force-induced degradation, the wet-and -dried powders can be milled or crushed by using a mortar and pestle.

When interpreting the results from the excipient compatibility study, the product development team should pay attention to the fact that the formula as a whole could behave differently from what may happen when each excipient alone reacts with the API molecule. Moreover, the results from excipient compatibility studies should not be and cannot be used as a quantitative basis to predict precisely how much degradation may occur during the required shelf life of the product.

Regarding shelf-life prediction, the Arrhenius equation is the only well-accepted model. However, the fundamental assumption in the Arrhenius theory is that the kinetic models for the reaction (degradation) of interest are the same at different temperatures. In reality, multiple mechanisms with different rate constants and activation energies often coexist, which render the Arrhenius equation not very useful. Even if the kinetics is simple, to make the extrapolation from the Arrhenius equation meaningful, the experimental results from analytical laboratories have to be accurate, which sometimes is difficult due to small/slow degradation. Therefore, the analytical scientists should keep in mind that instead of just saying avoid using certain excipients, the recommendation often is to use with caution. For example, from a pure chemistry perspective, it is definitely not compatible between acetylsalicylic acid and sodium carbonate or sodium bicarbonate.

However, there is a great product line of effervescent tablets that contain exactly aspirin and the carbonate salts. The pleasant effervescence of the solution provides a soothing sensation to those cough-and-cold sufferers and helps to relieve the symptoms. The shelf-life stability is achieved by manufacturing the effervescent products under extremely low humidity and by protecting the product with high-quality packaging materials to keep the moisture out.

5.7 Accelerated Stability Evaluation of Finished Products

The concept of implementing effective and efficient approaches in fast-paced product development repeatedly appear throughout this book. The desire to know the shelf-life stability of a new product after a few days or a few weeks rather than having to wait for three to six months after making the prototype is very much understandable. On the other hand, harsh approaches may generate irrelevant degradation, and mild approaches may take too long. To quickly predict future stability of a new formula, one can use commercially available software such as the "Accelerated Stability Assessment Program (*ASAP*®)." The predictions of the product shelf life are carried out based on stress studies on formulation prototypes for a period of a few weeks. Lots of literature reports can be found regarding the theory and practices of the program [73].

The author of this book had proposed another approach in 2009 [74]. The experimental design is based on the following thought process. First, since the worst condition that a finished product most likely will encounter during a stability study is 40°C/75% RH for six months (as shown in Table 5.1), stressing the prototype with a condition of (slightly) higher heat (less than 70°C) *plus* humidity can potentially shorten the time needed to predict the future stability. Second, due to the nature of OTC product development, it is quite possible that the new product contains a well-known API. There are products with similar dosage forms or API strengths already available on the market. In another word, the shelf life of the marketed product is already established. Therefore, the newly developed formula can be stressed side by side with some similar existing (commercial) products under the same conditions such as 50°C/75% RH or 60°C/75% RH for a few days to a few weeks in a noninvasive/nondestructive fashion. This approach can provide a quick read on the stability of the new formula by using an existing product as a reference for comparison. Figure 5.31 shows an example of such an experimental setup. The finished products, with or without packaging protection, are put in Petri dishes. Then, the Petri dishes are placed in a desiccator. The desiccator contains a saturated salt solution to maintain a designed relative humidity inside the desiccator. Various relative humidity can be created using different saturated

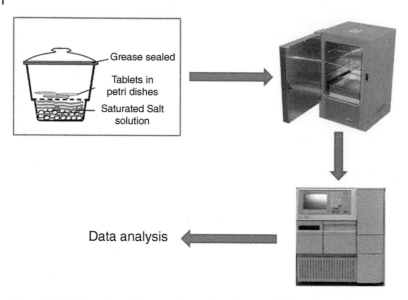

Figure 5.31 Experimental setup for conducting a predictive forced-degradation study for shelf-life estimation of a new formula.

Table 5.7 Examples of some saturated salt solutions and relative humidity.

Salt	25°C %RH	30°C %RH	40°C %RH	50°C %RH	60°C %RH
$MgCl_2$	40	40	40	40	40
NaBr	50	50	50	50	50
Nacl	75	75	75	75	75
KCl	85	85	85	80	80
$NaNO_2$	60	60	60	60	60
$NaNO_3$	65	65	65	65	65

salt solutions (Table 5.7). The desiccator is capped with grease seal and is then placed into an oven where the temperature is maintained at 50 or 60°C. At this point, since supposedly the API forced degradation and excipient compatibility studies have been performed, the analytical scientist should have a stability-indicating testing procedure in hand. More importantly, the analytical scientist knows which degradation peak(s) should be monitored. By plotting the chromatographic peak areas against the stressing time (days or weeks) for the new formula and the existing marketed product with similar formulas, the analytical scientist can conduct a semi-quantitative comparison. For example, Figure 5.32 contains three formulas. Formula A is a marketed product that has an established shelf life of two years. Formula B and formula C are two new products under development. Based on the product development knowledge, the formation of Degradant X is the most critical determining factor for the long-term stability of the formulas. As shown in Figure 5.32, the formation of Degradant X in formula B is similar to or even slightly lower than the formation of Degradant X in the marketed formula A. The conclusion from this observation is that formula B can potentially have a shelf life of two years, just like the marketed formula A. On the other hand, the formation of Degradant X in formula C is faster than the formation of Degradant X in formula A. Therefore, the long-term stability of formula C will be questioned.

There are several interesting findings from the above study. One is that various relative humidity can be created by mixing different salts to make saturated solutions. The readers can try to create various combinations of relative humidity and temperatures, if interested. For example, by mixing $NaNO_2$ and KCl in an appropriate ratio, a relative humidity of 70% can be created. Instead of using $NaNO_2$, a mixture of NaBr and KCl can create a 60% relative humidity. The other interesting finding is that radical-induced autoxidation stress can occur with the use of saturated nitrate or nitrite salt solution. The mechanism of this phenomenon was studied [75] by other scientists about ten years after the author of this book presented the finding in the 2009 AAPS Stability workshop.

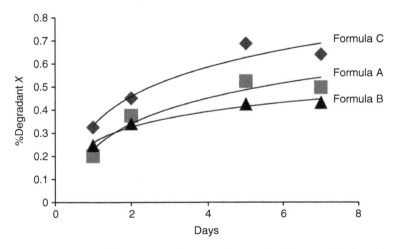

Figure 5.32 Plot of the growth of degradation product X versus the time (days). The plot is only for illustration purpose and does not represent any real data.

References

1 The U.S. Food and Drug Administration (FDA) (November 2003). Guidance for Industry Q1A(R2) Stability Testing of New Drug Substances and Products.

2 Loftsson, T. (2014). *Drug Stability for Pharmaceutical Scientists*. Elsevier Inc.

3 Li, M. (2012). *Organic Chemistry of Drug Degradation*. Royal Society of Chemistry.

4 Baertschi, S.W., Alsante, K.M., and Reed, R.A. (2016). *Pharmaceutical Stress Testing: Predicting Drug Degradation*, 2e. CRC Press.

5 Qiu, F. and Scrivens, G. (2018). *Accelerated Predictive Stability (APS): Fundamentals and Pharmaceutical Industry Practices*. Academic Press.

6 Huynh-Ba, K. (2009). *Pharmaceutical Stability Testing to Support Global Markets*. Springer Science & Business Media.

7 Teasdale, A., Elder, D., and Nims, R.W. (2017). *ICH Quality Guidelines: An Implementation Guide*. Wiley.

8 Narang, A.S. and Boddu, S.H.S. (2015). *Excipient Applications in Formulation Design and Drug Delivery*. Springer.

9 Al-Achi, A., Gupta, M.R., and Stagner, W.C. (2013). *Integrated Pharmaceutics: Applied Preformulation, Product Design, and Regulatory Science*. Wiley.

10 Brittain, H.G. (1996). *Analytical Profiles of Drug Substances and Excipients*. Academic Press.

11 Qiu, Y., Chen, Y., Zhang, G.G.Z. et al. (2009). *Developing Solid Oral Dosage Forms: Pharmaceutical Theory and Practice*. Academic Press.

12 Fersht, A.R. and Kirby, A.J. (1967). The hydrolysis of aspirin. intramolecular general base catalysis of ester hydrolysis. *Journal of the American Chemical Society* 89 (19): 4857–4863.

13 Edwards, L.J., Gore, D.N., Rapson, H.D.C., and Taylo, M.P. (1955). The hydrolysis of aspirin in pharmaceutical preparations. A limit test for free salicylic acid. *The Journal of Pharmacy and Pharmacology* 7: 892–904.

14 Moradlou, O. and Zare, M.A. (2014). Kinetic study of paracetamol hydrolysis: evaluating of the voltammetric data. *Analytical & Bioanalytical Electrochemistry* 6 (4): 450–460.

15 Koshy, K.T. and Lach, J.L. (1961). Stability of aqueous solutions of *N*-acetyl-*p*-aminophenol. *Journal of Pharmaceutical Sciences* 50 (2): 113–118.

16 Roehrig, R.C. (July 2011). CHPA's Proposal of Limit for the Process Degradant 4-Aminophenol.

17 Testa, B., Mayer, J.M., and Mayer, J. (2003). *Hydrolysis in Drug and Prodrug Metabolism*. Wiley.

18 Min, D.B. and Boff, J.M. (2002). Chemistry and reaction of singlet oxygen in foods. *Comprehensive Reviews in Food Science and Food Safety* 1: 58–72.

19 Hovorka, S.W. and Schöneich, C. (2001). Oxidative degradation of pharmaceuticals: theory, mechanisms and inhibition. *Journal of Pharmaceutical Sciences* 90 (3): 253–269.

20 Shilp, B., Anil, K., and Ameta, S.C. (2011). Photooxidation of some pharmaceutical drugs by singlet molecular oxygen. *Asian Journal of Pharmaceutical & Biological Research (AJPBR)* 1 (2): 210–217.

21 Frimer, A.A. (1979). The reaction of singlet oxygen with olefins: the question of mechanism. *Chemical Reviews* 79 (5): 359–387.

22 Boccardi, G. (1994). Autoxidation of drugs: prediction of degradation impurities from results of reaction with radical chain initiators. *Farmaco* 49 (6): 431–435.

23 Boccardi, G., Deleuze, C., Gachon, M. et al. (1992). Autoxidation of tetrazepam in tablets: prediction of degradation impurities from the oxidative behavior in solution. *Journal of Pharmaceutical Sciences* 81 (2): 183–185.

24 Watkins, M.A., Pitzenberger, S., and Harmon, P.A. (2013). Direct evidence of 2-cyano-2-propoxy radical activity during AIBN-based oxidative stress testing in acetonitrile–water solvent systems. *Journal of Pharmaceutical Sciences* 102 (3): 1554–1568.

25 Nelson, E.D., Thompson, G.M., Yao, Y. et al. (2009). Solvent effects on the AIBN forced degradation of cumene: implications for forced degradation practices. *Journal of Pharmaceutical Sciences* 98 (3): 959–969.

26 Salgado, P., Melin, V., Contreras, D. et al. (2013). Fenton reaction driven by iron ligands. *Journal of the Chilean Chemical Society* 58 (4).

27 Yu, H., Cornett, C., Larsen, J., and Hansenb, S.H. (2010). Reaction between drug substances and pharmaceutical excipients: formation of esters between cetirizine and polyols. *Journal of Pharmaceutical and Biomedical Analysis* 53 (3): 745–750.

28 Sharma, N., Rao, S.S., Ullas, G. et al. (2011). Optimization and validation of reverse phase liquid chromatographic method for estimation of cetirizine mannitol ester impurity in cetirizine hydrochloride chewable tablet. *Journal of Liquid Chromatography & Related Technologies* 34: 2157–2168.

29 Liu, L., Martellucci, S., Xiao, K.P. et al. (2010). Investigation into the formation of ibuprofen methyl ester in aqueous methanol solutions of different pH. *American Pharmaceutical Review* 13 (5): 54–59.

30 Troup, A.E. and Mitchner, H. (1964). Degradation of phenylephrine hydrochloride in tablet formulations containing aspirin. *Journal of Pharmaceutical Sciences* 53 (4): 375–379.

31 Aulton, M.E. and Taylor, K.M.G. (2017). *Aulton's Pharmaceutics E-Book: The Design and Manufacture of Medicines*. Elsevier Health Sciences.

32 Ma, M., DiLollo, A., Mercuri, R. et al. (2002). HPLC and LC–MS studies of the transesterification reaction of methylparaben with twelve 3- to 6-carbon sugar alcohols and propylene glycol and the isomerization of the reaction products by acyl migration. *Journal of Chromatographic Science* 40: 170–177.

33 Farsaa, O., Subertb, J., and Marečkováa, M. (2011). Hydrolysis and transesterification of parabens in an aqueous solution in the presence of glycerol and boric acid. *Journal of Excipients and Food Chemistry* 2: 41–49.

34 Ma, M., Lee, T., and Kwong, E. (2002). Interaction of methylparaben preservative with selected sugars and sugar alcohols. *Journal of Pharmaceutical Sciences* 91: 1715–1723.

35 Fan, G., Zhan, J., Luo, J. et al. (2019). Photocatalytic degradation of naproxen by a $H_2O_2^-$ modified titanate nanomaterial under visible light irradiation. *Catalysis Science & Technology* 9: 4614–4628.

36 Kumar, S. and Burgess, D.J. (2014). Wet milling induced physical and chemical instabilities of naproxen nano-crystalline suspensions. *International Journal of Pharmaceutics* 466 (1–2): 223–232.

37 Trommer, H., Raith, K., and Neubert, R.H.H. (2010). Investigating the degradation of the sympathomimetic drug phenylephrine byelectrospray ionisation–mass spectrometry. *Journal of Pharmaceutical and Biomedical Analysis* 52: 203–209.

38 Douša, M., Gibala, P., Havlíček, J. et al. (2011). Drug–excipient compatibility testing – identification and characterization of degradation products of phenylephrine in several pharmaceutical formulations against the common cold. *Journal of Pharmaceutical and Biomedical Analysis* 55: 949–956.

39 Maillard, L.C. (1912). Action of amino acids on sugars. Formation of melanoidins in a methodical way. *Comptes Rendus de l'Académie des Sciences* 154: 66–68.

40 Marin, A., Espada, A., Vidal, P., and Barbas, C. (2005). Major degradation product identified in several pharmaceutical formulations against the common cold. *Analytical Chemistry* 77: 471–477.

41 Wong, J., Wiseman, L., Al-Mamoon, S. et al. (2006). Correspondence major degradation product identified in several pharmaceutical formulations against the common cold. *Analytical Chemistry* 78: 7891–7895.

42 Blessy, M., Ruchi, D.P., Prajesh, N.P., and Agrawa, Y.K. (2014). Development of forced degradation and stability indicating studies of drugs – a review. *Journal of Pharmaceutical and Biomedical Analysis* 4 (3): 159–165.

43 Alsante, K.M., Ando, A., Brown, R. et al. (2007). The role of degradant profiling in active pharmaceutical ingredients and drug products. *Advanced Drug Delivery Reviews* 59: 29–37.

44 Shinde, N.C., Bangar, B.N., Deshmukh, S.M. et al. (2013). Pharmaceutical forced degradation studies with regulatory consideration. *Asian Journal of Research in Pharmaceutical Sciences* 3 (4): 178–188.

45 Nelson, E.D., Harmon, P.A., Szymanik, R.C. et al. (2006). Evaluation of solution oxygenation requirements for azonitrile-based oxidative forced degradation studies of pharmaceutical compounds. *Journal of Pharmaceutical Sciences* 95 (7): 1527–1539.

46 Jain, D. and Basniwala, P.K. (2013). Forced degradation and impurity profiling: recent trends in analytical perspectives. *Journal of Pharmaceutical and Biomedical Analysis* 86: 11–35.

47 Kovaleski, J., Kraut, B., Mattiuz, A. et al. (2007). Impurities in generic pharmaceutical development. *Advanced Drug Delivery Reviews* 59: 56–63.

48 Thatcher, S.R., Mansfield, R.K., Miller, R.B. et al. (2001). Pharmaceutical photostability a technical guide and practical interpretation of the ICH guideline and its application to pharmaceutical stability – part II. *Pharmaceutical Technology* 25: 50–62.

49 Dave, V.S., Saoji, S.D., Raut, N.A., and Haware, R.V. (2015). Excipient variability and its impact on dosage form functionality. *Journal of Pharmaceutical Sciences* 104 (3): 906–915.

50 Wu, Y., Levons, J., Narang, A.S. et al. (2011). Reactive impurities in excipients: profiling, identification and mitigation of drug–excipient incompatibility. *AAPS PharmSciTech* 12 (4): 1248–1263.

51 Wasylaschuk, W.R., Harmon, P.A., Wagner, G. et al. (2007). Evaluation of hydro-peroxides in common pharmaceutical excipients. *Journal of Pharmaceutical Sciences* 96 (1): 106–116.

52 Pan, C., Liu, F., and Motto, M. (2011). Identification of pharmaceutical impurities in formulated dosage forms. *Journal of Pharmaceutical Sciences* 100 (4): 1228–1259.

53 Soh, J.L.P., Liew, C.V., and Heng, P.W.S. (2015). Impact Of excipient vriability on drug product processing and performance. *Current Pharmaceutical Design* 21 (40): 5890–5899.

54 Narang, A.S., Desai, D., and Badawy, S. (2012). Impact of excipient interactions on solid dosage form stability. *Pharmaceutical Research* 29 (10): 2660–2683.

55 Li, G., Schoneker, D., Ulman, K.L. et al. (2015). Elemental impurities in pharmaceutical excipients. *Journal of Pharmaceutical Sciences* 104 (12): 4197–4206.

56 Boetzel, R., Ceszlak, A., Day, C. et al. (2018). An elemental impurities excipient database: a viable tool for ICH Q3D Drug Product Risk Assessment. *Journal of Pharmaceutical Sciences* 107 (9): 2335–2340.

57 Szakonyi, G. and Zelkó, R. (2012). The effect of water on the solid state characteristics of pharmaceutical excipients: molecular mechanisms, measurement techniques, and quality aspects of final dosage form. *International Journal of Pharmaceutical Investigation* 2 (1): 18–25.

58 Ohtake, S. and Shalaev, E. (2013). Effect of water on the chemical stability of amorphous pharmaceuticals: I. Small molecules. *Journal of Pharmaceutical Sciences* 102 (4): 1139–1154.

59 Hancock, B.C. and Zografi, G. (1994). The relationship between the glass transition temperature and the water content of amorphous pharmaceutical solids. *Pharmaceutical Research* 11 (4): 471–477.

60 Mehta, M., Kothari, K., Ragoonanan, V., and Suryanarayanan, R. (2016). Effect of water on molecular mobility and physical stability of amorphous pharmaceuticals. *Molecular Pharmaceutics* 13 (4): 1339–1346.

61 Ahlneck, C. and Alderborn, G. (1988). Solid state stability of acetylsalicylic acid in binary mixtures with microcrystalline and microfine cellulose. *Acta Pharmaceutica Suecica* 25 (1): 41–52.

62 Pudipeddi, M., Zannou, E.A., Vasanthavada, M. et al. (2008). Measurement of surface pH of pharmaceutical solids: a critical evaluation of indicator dye-sorption method and its comparison with slurry pH method. *Journal of Pharmaceutical Sciences* 97 (5): 1831–1842.

63 Li, J. and Wu, Y. (2014). Lubricants in pharmaceutical solid dosage forms. *Lubricants* 2: 21–43.

64 Fathima, N., Mamatha, T., Qureshi, H.K. et al. (2011). Drug–excipient interaction and its importance in dosage form development. *Journal of Applied Pharmaceutical Sciences* 1 (6): 66–71.

65 Holm, R. and Elder, D.P. (2016). Analytical advances in pharmaceutical impurity profiling. *European Journal of Pharmaceutical Sciences* 87 (25): 118–135.

66 Sims, J.L., Carreira, J.A., Carrier, D.J. et al. (2003). A new approach to accelerated drug–excipient compatibility testing. *Pharmaceutical Development and Technology* 8 (2): 119–126.

67 Wyttenbach, N., Birringer, C., Alsenz, J., and Kuentz, M. (2005). Drug–excipient compatibility testing using a high-throughput approach and statistical design. *Pharmaceutical Development and Technology* 10 (4): 499–505.

68 Serajuddin, A.T.M., Thakur, A.B., Ghoshal, R.N. et al. (1999). Selection of solid dosage form composition through drug–excipient compatibility testing. *Journal of Pharmaceutical Sciences* 88 (7): 696–704.

69 Raijada, D., Müllertz, A., Cornett, C. et al. (2014). Miniaturized approach for excipient selection during the development of oral solid dosage form. *Journal of Pharmaceutical Sciences* 103 (3): 900–908.

70 Thomas, V.H. and Naath, M. (2008). Design and utilization of the drug–excipient chemical compatibility automated system. *International Journal of Pharmaceutics* 359 (1–2): 150–157.

71 Chadha, R. and Bhandari, S. (2014). Drug–excipient compatibility screening – role of thermoanalytical and spectroscopic techniques. *Journal of Pharmaceutical and Biomedical Analysis* 87 (18): 82–97.

72 Monkhouse, D.C. and Maderich, A. (1989). Whither compatibility testing? *Drug Development and Industrial Pharmacy* 15 (13): 2115–2130.

73 Waterman, K.C. (2011). The application of the accelerated stability assessment program (ASAP) to quality by design (QbD) for drug product stability. *AAPS PharmSciTech* 12: 932.

74 Xiao, K. P. and Gentry, A. (2009). Challenges and case studies with developing predictive forced-degradation models for finished dosage forms. *AAPS Workshop on Current Trends in Stability – Challenges with Today's New Products* (24–25 September 2009) Gaylord National Resort & Convention Center National Harbor, MD.

75 Sluggetta, G.W., Zeleskya, T., Hetrickb, E.M. et al. (2018). Artifactual degradation of secondary amine-containing drugs during accelerated stability testing when saturated sodium nitrite solutions are used for humidity control. *Journal of Pharmaceutical and Biomedical Analysis* 149: 206–213.

6

Practical Statistics for Analytical Development

In the pharmaceutical industry, statistics seem to be more associated with clinical trial designs and for analyzing study outcomes [1–5]. Although modern clinical trials are so-called evidence-based, appropriate statistical models are the foundations for correctly demonstrating or rejecting the hypothesis behind the study. Professional statisticians are critical team members in clinical trials. Recently, however, as a trend, the health authorities, pharmacopeia bodies, and academia scholars are pushing the pharmaceutical industry to take statistical approaches also for analytical data analysis, risk assessment, stability data trend analysis, method validation and transfer, method comparison, and method life cycle management. Analytical target profile (ATP), target measurement uncertainty (TMU), quality by design (QbD), design of experiments (DoE), and analytical control strategy (ACS), are a few examples of those popular topics often discussed within the analytical chemistry society [6–11]. On the other hand, there are lots of nuances, details, and hidden assumptions in statistical analyses. The reality is that an analytical scientist is usually not a statistician. It is not an exaggeration to say the understanding of the assumptions behind the statistical approaches, the grasp of the theoretical or practical meaning of obtained test statistics, are lacking in many analytical laboratories. For example, even a seemingly simple concept such as *average* can be misused. If you were given a question like the following: "A person drives through a residential area at a speed of 20 kilometers/h, followed by driving on a highway at a speed of 100 kilometers/h. What is the average speed that this person drives?" you would immediately know that you cannot just add 100 and 20 then divide the sum of 120 by 2, and say the average speed is 60 km/h. That is because *kilometer* and *hour* are two independent variables, and when dividing the sum of 120 km/h by 2, the unit of 2 is unclear. You would ask for more information, such as how much time the person drives in the residential area, and how much time the person drives on the highway. Based on that information, you would calculate the entire distance (kilometers) the person drives and

Analytical Scientists in Pharmaceutical Product Development: Task Management and Practical Knowledge, First Edition. Kangping Xiao.
© 2021 John Wiley & Sons, Inc. Published 2021 by John Wiley & Sons, Inc.

divides that by the total driving time (hours), the result is then the average speed of driving. That seems simple and straightforward. However, what if you were asked whether you could average ten content uniformity (CU) test results and report the mean as the potency of the product, instead of conducting an assay testing that is usually performed by analyzing ten units together as one sample? You may find some companies take that approach as a timesaving innovation, although the practice of adding up 10 CU results and then dividing the sum by 10, is fundamentally false. Interestingly, even the USP allows that approach under some conditions that are not clearly defined. Section 5.70 of the General Notices says that when the same analytical methodology is used for the assay and the content uniformity with appropriate allowances made for differences in sample preparation, the average of all the individual-unit CU determinations may be used as the assay value [12]. A closer look at the results will reveal that for each testing result obtained from every single unit, the data has two variables associated with the value, one is the random product variation, and the other one is the inherent analytical method variation. Those two variables are independent of each other, which renders the "averaging CU results" approach less meaningful. The denominator, 10, loses its hidden unit. It is not well defined whether it is the 10 products, so to average product variability, or it should be 10 analyses, so to average analytical measurement variability. It is different from averaging ten chromatographic injections from the same sample solution, where the hidden unit of the denominator 10 is the injection. Therefore, it is highly recommended that the analytical scientists work with professional statisticians or at least consult with the experts when conducting some statistical analyses.

Have you heard this saying: "Lies, damned lies, and statistics"? It tells us how much power the statistics and numbers can have on people. Indeed, when we hear things like, it is statistically proven that Brand X medicine works better than does the Brand Y medicine of the same kind, consciously or unconsciously, our brain will make us believe that claim. Therefore, it can be beneficial to an analytical scientist if he or she has more than cursory knowledge of statistics. This chapter aims at describing some statistical tools and their relevancy to analytical work. The emphasis is to translate the statistical language into the language a nonstatistician analytical scientist can comprehend. This chapter only describes rather basic concepts at elementary levels. The purpose is to make the analytical scientists aware of those basic statistical terms and applications, so they can have educated conversations with professional statisticians, or have meaningful discussions among themselves when encountering statistics. What is more important, though, is that analytical scientists should remember that the statistics are only as good as the data that feed the models. The designs of the laboratory experiments or studies have to be scientifically reliable for the statisticians to work on meaningful data. Moreover, analytical scientists should be able to judge the practical meaning of statistical differences. For example, we all know that sample sizes are critical in statistical calculations. We tend to make sure the sample sizes are sufficiently large to ensure the statistical outcomes are sound and meaningful. However, we may not know that when the sample sizes are large, statistical tests tent to show that small differences are statistically significant, even though those differences may not have any practical significance. The analytical scientists need to make a correct judgment when finding some conclusion is statistically significant but analytically irrelevant [13, 14].

Readers who are interested in more sophisticated statistics can further move on to other specialized books or literature for further study.

6.1 Basic Statistical Terms

6.1.1 Sample Versus Population

The word *sample* used in statistical language is not a sample that an analytical scientist is working on daily, such as a tablet or a solution prepared for assay, dissolution, or degradation product analysis. The word *sample* as a statistical term describes a pool of data obtained from analyzing the analytical testing samples. To differentiate, we will use "statistical sample" for statistical purposes and use the phrase "analytical sample" to describe a sample used for laboratory work. By evaluating the entirety of the data and each data point individually in one statistical sample, scientists wish to come up with a specific claim or discover a particular pattern of data distribution in a *population*. In other words, scientists hope to be able to describe a fact discovered regarding a large amount of data (*population*) through the analysis of a limited amount of data (statistical sample). For example, when evaluating a tablet manufacturing process, the *population* can be the abundant historical data accumulated throughout the years since the establishment of the manufacturing process or can be the data generated from hundreds and thousands of tablets made during one production campaign. When evaluating an analytical method, the *population* can be the existing large amount of data points generated by the method historically, plus the future data that yet to be produced by the use of this analytical method in one laboratory.

This sample versus population concept is essential for analytical scientists to always keep in mind when attempting to use statistics for data evaluation. Some unintelligent conclusions can be made when mixing up those two concepts, especially in a R&D analytical laboratory where the size of a statistical sample, i.e. the number of data points obtained from completely independent experiments, is usually small. Whether or not the statistical sample can adequately represent the population should be the first question the analytical scientist asks. Moreover, when comparing averages of or variability in two sets of statistical samples, analytical scientists should know that it is not just to compare the numerical results from those two sets of analytical samples. The real goal is to statistically compare means of or variability in two sets of populations that are represented by those two sets of statistical samples.

6.1.2 Mean, Variance, Standard Deviation, and Relative Standard Deviation

In our daily life or work, we like to use the word "average." In the statistical world, the word "mean" is used much more often. A mean of a statistical sample is the ratio between the sum of all the data (x_1, x_2, ..., x_n) and the number of measurements (n). The equation is shown below:

$$\bar{x} = \frac{\sum x_i}{n} = \frac{x_1 + x_2 + \cdots + x_n}{n} \tag{6.1}$$

A "mean" is different from another statistical term, "median." A median is a data or result that is in the middle of measurements. For example, if the number of measurements is 5 (an odd number), and they are in the order of "measurement 1," "measurement 2," "measurement 3," "measurement 4," and "measurement 5," then the "measurement 3" is the middle number of those five measurements. The data from that "measurement 3" is then the median from those five measurements. If the number of measurements is 6 (an even number), then "measurement 3" and "measurement 4" are both located in the middle, and the average of the data from "measurement 3" and "measurement 4" is then the median in those six measurements. Another term frequently used to describe some sort of general behavior in a population is "mode." Mode is the value that occurs most often, and if there is no repeated data in the statistical sample, then there is no mode for that sample.

A variance quantitatively describes the scatter of the data in one statistical sample or one population. To calculate the variance, the distance between each data and the mean of a statistical sample ($\bar{x}$), or the distance between each data and the mean of a population (μ), is squared. The sum of those squares is divided either by the degree of freedom ($n - 1$, number of measurements minus one) to obtain the variance of the statistical sample (s^2) or is divided by the number of measurements (n) to obtain the variance of the population (σ^2). By squaring the distances, the variance prevents the cancellation of the positive and negative differences of the data from the mean. The equations are shown below:

$$s^2 = \frac{\sum_i \left(x_i - \bar{x}\right)^2}{n-1} \quad \text{Statistical sample variance} \tag{6.2}$$

$$\sigma^2 = \frac{\sum_i \left(x_i - \mu\right)^2}{n} \quad \text{Population variance} \tag{6.3}$$

More popular than variance, standard deviation (SD) quantitatively describes the variation or spread of the values in one statistical sample or one population. The smaller the SD is, the closer each data is to the mean. The equations are shown below:

$$s = \sqrt{s^2} \quad \text{Statistical sample standard deviation} \tag{6.4}$$

$$\sigma = \sqrt{\sigma^2} \quad \text{Population standard deviation} \tag{6.5}$$

The SD is more tangible than variance for variation evaluation as it has the same unit as the original data of measurement. For example, if the mean is 100 mg, the SD is also expressed in mg. If one SD is ±10 mg, it gives us a sense of what is the *normal* distance to the mean, and what is deviating maybe slightly too much, i.e. between 90 and 110 mg, or between 80 and 120 mg. In this example, ±20 mg represents a two SD, and ±30 mg represents a three SD (70–130 mg). Depending on how many SDs are acceptable, the *normal* versus *abnormal* characteristics of a behavior, a performance, a measurement, etc., can be described quantitatively.

Now here comes the most popular statistical term, the relative standard deviation (RSD), also known as coefficient of variation (CV). It is the ratio of the SD over the mean, expressed in percentage. It is extensively used in analytical laboratories to express the precision and repeatability of a technique or a set of measurements. The equations are shown below:

$$\%\text{RSD} = \text{CV} \times 100\% = \frac{s}{\bar{x}} \times 100\% \quad \text{Statistical sample relative standard deviation} \tag{6.6}$$

$$\%\text{RSD} = \frac{\sigma}{\mu} \times 100\% \quad \text{Population relative standard deviation} \tag{6.7}$$

The %RSD provides another perspective, in a relative term, on assessing the variation in a statistical sample or a population. As a unitless value, %RSD can be used to compare results derived from different categories of measurements, such as comparison between speed (m/s) and density (g/L). It was said that the first use of %RSD was to compare men and women in 1895, by the biostatistician Karl Pearson to calculate the averages and SDs of male and female internal organs. The organ sizes of men were larger and had higher variability. To compare them on the same basis, he adjusted the SD by dividing it by the average, multiplying by 100, then yielded a percentage.

However, although this seemingly simple tool has found broad acceptance in statistical analyses, misuses of %RSD in analytical laboratories are not uncommon.

The unitless nature of %RSD provides a universal base for comparison, but at the same time can make people forget the fundamental question of "percent of what?" The most common mistake that analytical scientists make when using %RSD is to use it to communicate the recovery data that contain results from different concentration levels. Take Table 6.1 as an example: using the entire data set in Table 6.1 to calculate the %RSD can lead to a completely misleading overall conclusion on the method accuracy and precision. An accurate mean recovery of 99.4% together with a modest RSD of 1.8%, paint a picture of a good method. Looking closer, we can see that the recovery at the most critical concentration level, i.e. the 100% nominal level, is 2.4% higher than the theoretical 100% recovery, and the %RSD is a stunning 6.3%, which paints a very different picture of the method. Therefore, the question "percent of what" should be asked before performing Excel spreadsheet calculations. The %RSD of recoveries should be separately evaluated for each concentration level. To scientifically report the results from the above table, we need to present at least the following related information at each concentration, such as mean recovery, %RSD, and the number of replicates. Therefore, to report the results from Table 6.1, we will say that the analytical recovery at 100% nominal concentration level is 102.4% with a %RSD of 6.3%, based on three replicates; the recovery at 125% theoretical concentration level is 97.7% with a %RSD of 2.1%, based on three replicates, and so on. Another easy-to-forget point of the %RSD is that, by definition, a %RSD involves only one SD (see Eq. (6.6) and notice that the numerator is only one s). As we will see later, that one SD means that the confidence level associated with the reported result is only 68% at best. Therefore, when analyzing data with a finite number of a statistical sample, the confidence level we have regarding the behavior or trend

Table 6.1 Example of misuse of %RSD.

Accuracy/recovery of compound A and result precision

Concentration levels (percent of nominal)	Recovery (%) ($n = 3$)	Mean recovery (%)/ RSD (%) at each level
50	99.2, 98.6, 100.1	99.3/0.8
75	98.4, 96.2, 100.3	98.3/2.1
100	95.5, 108.2, 103.4	102.4/6.3
125	99.2, 98.6, 95.3	97.7/2.1
150	99.5, 99.2, 99.7	99.5/0.3

Mean recovery: 99.4%; overall RSD 1.8%

of the underlying population that the statistical sample is representing can be less than 68%, a refresh of memory the discussion of Sample versus Population. Indeed, since the calculation of variance is affected by the statistical sample size, estimates of the SD, and thus the %RSD can have enormous variations for small sample sizes. The analytical scientists should keep in mind that a calculated SD or %RSD during a method validation or transfer is not a fixed value. Different statistical samples may yield different values for both. It is a good way to look at this fact as an indication of the importance of the analytical method life cycle management. We should not be surprised when encountering issues with methods that have already been validated.

In addition, caution is needed when interpreting the %RSD obtained from data with a very small mean. As can be seen from Eq. (6.6), when the mean becomes smaller, the %RSD gets larger.

6.1.3 Normal Distribution

With the mean, variance, SD, and %RSD in mind, we now know that a statistical sample consists of a pool of data points that distribute around their mean, with some degree of variance/SD in the data. If we think more abstractly, by replacing the word "data points" with "probability," and substituting the word "mean" with "statement" or "hypothesis," then we can treat a statistical analysis as a tool to describe a distribution of a probability regarding the trueness of a statement or how sound a hypothesis is. This mental exercise helps us to understand why statistics are used so widely, almost in every field of scientific research, finance, social studies, or even business management.

The most common probability distribution found in nature is the so-called *Normal Distribution*, also called *Gaussian Distribution*, which is named after the mathematician Karl Friedrich Gauss. The fact is, though, it was Abraham de Moivre, an eighteenth century statistician who first discovered the normal distribution. The shape of the distribution somewhat resembles a bell, and thus the *Bell Curve* is a worldly name for Gaussian distribution. Some important features of a normal distribution (Figure 6.1) include: (1) The shape of the curve is symmetric around a mean (or a statement, a hypothesis). (2) The distribution is characterized by the mean ($\bar{x}$ or μ) and the SD (s or σ). (3) Approximately 68, 95, and 99.7% of the area under the normal distribution curve is within one SD of the mean $[\bar{x} - s, \bar{x} + s]$ or $[\mu - \sigma, \mu + \sigma]$, two SDs of the mean $[\bar{x} - 2s, \bar{x} + 2s]$ or $[\mu - 2\sigma, \mu + 2\sigma]$, and three SDs of the mean $[\bar{x} - 3s, \bar{x} + 3s]$ or $[\mu - 3\sigma, \mu + 3\sigma]$, respectively. Astonishingly, the famous Six Sigma approach widely used for process improvement is asking for a production control that the lower and upper specification limit is each 6σ away from the mean of the specification limits! For example, if a

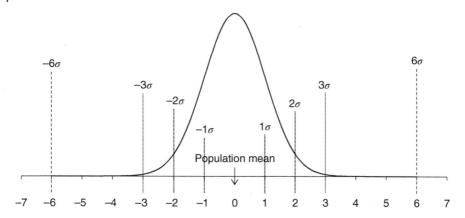

Figure 6.1 A normal distribution curve with a population mean and standard deviations.

specification of a tablet hardness is 10 ± 2 kp, with the lower limit being 8 kp (left side of the curve) and the upper limit being 12 kp (right side of the curve), the Six Sigma approach requires the compression of the tablets to be controlled to make 8 kp be -6σ away from 10 kp (the mean), and 12 kp be $+6\sigma$ away from 10 kp. In plain English, it means that the 6σ requires that the tablet compression process must be robust enough not only to guarantee that all the tablets (99.7% as the 3σ covers) produced are within the specification limits, but also buffers that guarantee with an extra 3σ SDs (Figure 6.1).

A pharmaceutical analytical scientist sees this normal distribution almost daily, that is, the shape of a chromatographic peak. Although on the chromatogram, the x-axis is time and the y-axis is usually the UV response, we can mentally switch the y-axis from UV intensity to probability intensity. The word "probability" here refers to the chance of observing the analyte compound of interest; the word "mean" here corresponds to the retention time of the peak for the analyte compound on the chromatogram (assuming a good chromatography, i.e. symmetrical peak shape). The chromatographic peak thus illustrates the distribution of the probability of that analyte compound elutes at that retention time. The more efficient the column is (i.e. with higher chromatographic column plate numbers), the more probability of observing that analyte molecules "showing up" at the labeled retention time. Therefore, the peak gets slimmer and taller as the analyte molecules all tend to elute at the same retention time. It can be a useful insight that a chromatographic peak reflects a statistical phenomenon. During one injection, the molecules of the analyte compound distribute themselves by centering on the suppose-to-be retention time. If the column efficiency is low or an analyte concentration is too high, the SD of the distribution gets larger. The observed chromatographic behavior is that the peak of analyte gets broader, and the tails of the peaks stretch to one or both ends, far away from its labeled retention time, and even never touch the baseline. When the peak broadens, it may stretch to the retention region of other compounds and thus interfere with the quantitation. This is especially true in the case of a sample that contains a very concentrated API. Since the concentration of that API is so high compared to other related compounds, even if there appears to be no co-elution on the chromatograms to our naked eyes, those related compound molecules may still "bump into" the API molecules far away before the front tail or far away after the end tail of the API peak. In such cases, the accuracy of the quantitation of those related compounds will suffer, and the API itself becomes the matrix in the sample solution, as far as the related compounds are concerned.

6.1.4 *t*-Statistics, *F*-Test, and ANOVA

The *t*-statistics was first published in 1908 by William Sealy Gosset, who used "Student" as the pen name in his paper. The *t* is designed to compare two means of two statistical samples, or to compare the mean of one statistical sample to a true value or a known number that can be practically considered as the mean of a population. The comparison is to see whether those two values/two means are _statistically different_. The *t*-statistics was introduced to conduct statistical analyses on samples with limited sizes. The distribution of the limited number of data points follows *t*-distribution, which is similar to the normal distribution.

When comparing the mean of one set of data to a known value, it should be kept in mind that this value is not some random number. It should be a well-known number that is considered as a true value, a golden standard. Good examples of such values are the standards provided by the National Institute of Standards and Technology (NIST). The laboratories declare the balance weights they use are NIST traceable since the calibration of laboratory balance weights is measured against the NIST standards. In statistical language, this NIST weight standard is the "True Statement," or "True Value," or "Population Mean," etc. Another example of such a number is a specification limit, which is more familiar to the pharmaceutical analytical scientists. The calculation of such *t* is shown below:

$$t = \frac{\overline{x} - \mu}{s_x} \times \sqrt{n} \tag{6.8}$$

where $\bar{x}$ is the mean of the statistical sample and the μ is the known value, s_x is the SD of the statistical sample, and n is the number of data points. The above equation is most of the time expressed in the following way:

$$t = \frac{\bar{x} - \mu}{s_x / \sqrt{n}} \tag{6.9}$$

where $s_x / \sqrt{n}$ is called standard error (SE), which is the average of the SD of all the data points in the statistical sample.

Quite disturbingly, however, there seem to be many incorrect uses of statistical models and mixing up of statistical concepts in quite a large portion of published clinical studies. It has been reported in numerous literature or web blogs that even the basic statistical concepts, such as those SD and SE, are frequently misused. In statistics, if normally distributed, the study sample can be described entirely by two parameters: the mean and the SD. The SD represents the variability within the sample. The SE is used in inferential statistics to give an estimate of how the mean of a particular statistical sample is related to the (true) mean of the underlying population ($SE = SD / \sqrt{n}$). More specifically, SE estimates the precision and uncertainty of how the study sample represents the underlying population. As the SE is always smaller than the SD, when the within-sample variations are presented by the SE instead of the SD, the nonstatistician readers may think that the variability within the sample is much smaller than it really is. Stunningly, based on an article published by Nagele [15], in 2001 alone, around 20–30% of anesthesia journals published articles that practiced the concept of SE incorrectly.

Equations (6.8) and (6.9) apply to one-sample t-tests. The assumption behind the mathematics is that the data being evaluated should be randomly selected and are independent of each other. The data is approximately normally distributed. Remember, again, this normal distribution requirement is referring to the population that the statistical sample represents, and not referring to the statistical sample itself. Think about the fact that 6–12 analytical samples are the most common numbers that analytical laboratories use to conduct experimental testing. A requirement of a normal distribution for every 6–12 data points is not practical, to say the least.

When comparing the means from two statistical samples, the statistics becomes a two-sample t-test. The numerator in Eq. (6.10) is the difference between the two means ($\bar{x} - \bar{y}$), while the denominator has some variations depending on the situations. The original Student's t-test for the two-samples t-test has the assumption that the variances in the data of the two statistical samples are statistically undistinguishable, in addition to the default assumption that the data in both statistical samples follow the normal distribution. To make it very clear, let us say it again that it does not mean at all that the two variances are exactly the same numerical numbers; it just means that there is no statistical difference in the two variances. The equation for calculating such a t is shown below.

$$t = \frac{\bar{x} - \bar{y}}{s_p\sqrt{\dfrac{1}{n_x} + \dfrac{1}{n_y}}}$$

(6.10)

$$s_p = \sqrt{\frac{(n_x - 1)s_x^2 + (n_y - 1)s_y^2}{n_x + n_y - 2}}$$

where s_p is a pooled SD of the two statistical samples; s_x and s_y is the SD of statistical sample x and statistical sample y, respectively; n_x and n_y are the number of data points in each statistical sample.

As an evolution of the original application, the t test is also used when the two variances are not statistically equal, in which case the calculation is sometimes referred to as Welch's t-test. The equation is as follows:

$$t = \frac{\bar{x} - \bar{y}}{\sqrt{\dfrac{s_x^2}{n_x} + \dfrac{s_y^2}{n_y}}}$$

(6.11)

A variation of the one-sample t-test is the paired sample t-test. This t-test is to compare the average of the differences in each pair of the results obtained on individual analytical samples, to a true average of differences, which is zero. For example, let us look at the injection precision of an HPLC by injecting each sample solution twice. In an ideal world, the peak areas from each couple of replicate injections are identical. That means the variance in peak areas between each pair of duplicate injections is zero. Therefore, the mean of all the differences is also zero. In the real world, however, there is always some difference in the duplicate injections. In this case, the statistical samples (the data points) are the differences between each pair of duplicate injections. The mean in this statistical sample is the average of all those observed differences. Since the μ is 0, the equation becomes:

$$t = \frac{\bar{D} - \mu}{s_D} \times \sqrt{n} = \frac{\bar{D} - 0}{s_D} \times \sqrt{n} = \frac{\bar{D}}{s_D} \times \sqrt{n}$$

(6.12)

with $D = x_i - y_i$

$$\bar{D} = \frac{\sum(x_i - y_i)}{n}$$

where D represents the difference between each pair of the data, $\bar{D}$ represents the mean of the differences; s_D is the SD of the differences, and n is the number of paired samples. The purpose of the test is to determine whether there is statistical evidence that the mean difference between paired observations is statistically meaningful, i.e. whether the mean difference is *significantly different* from zero.

When the *t*-test was originally proposed by "Student," it was introduced to handle sample size less than, for example, 30. There is another test, a *z-test*, that is designed to computing larger sample sizes. The calculation is shown below:

$$Z = \frac{\bar{x} - \mu}{\sigma} \times \sqrt{n} \qquad (6.13)$$

It can be seen that the *z*-test is a *t*-test with the use of population mean and population SD. Numerically, the larger the sample size, the closer a *t* distribution follows a normal distribution, which is represented by the statistics of *z*, and the influence of the degrees of freedom (i.e. the difference between n and $n - 1$) on the shape of distribution gradually becomes insignificant.

However, it is very common to find that a *t*-test is used by statisticians to compute the statistical analyses for sample sizes much larger than 30. With nowadays computer calculation power, many statistical packages use a *t*-test regardless of whether the sample size is small or large. In reality, since it is hardly ever for a pharmaceutical analytical scientist to be able to obtain hundreds and thousands of data to compute the population variance, it is rational and practical to use a *t*-test in our work.

As we already know, to describe a normal distribution, we not only need to know the mean but also the variance/SD of the data. To compare whether the scatterings of the data from their means in each distribution are *statistically equal*, we will use an *F*-test. The character "F" is named in honor of Ronald Fisher, one of the twentieth century's greatest scientists, who was knighted by Queen Elizabeth and became Sir Ronald Aylmer Fisher in 1952. If we start to study the *F*-test from the *F*-distribution, we may get ourselves into too much of the statistics. Fortunately as an analytical scientist, we may only need to know that an *F*-test is used to compare the two variances in two data distributions, and the *F*-value is simply the ratio of those two variances (the squares of corresponding SDs), as shown below:

$$F = \frac{s_x^2}{s_y^2} \qquad (6.14)$$

To execute a *t*-test or an *F*-test, we compare the calculated *t*-value or *F*-value with the values found in *t*-tables or in *F*-tables, which can be easily found in textbooks or on the Internet. If the calculated *t*-value is larger than the corresponding critical value listed on the *t*-table, we can say the two means are *statistically different*. If the calculated *F*-value is larger than the corresponding critical value on the *F*-table, we can say the two variances are *not statistically equal*.

So far we have described the statistical comparisons between two means for two groups, or between one group and one population. When we are trying to compare whether the means from two or more groups are statistically equal, we perform an analysis of variance (ANOVA). However, different from the Student's *t*-test, ANOVA actually compares the *variances* in the groups, rather than the *means* themselves. It was first proposed by Sir Ronald Fisher in 1925. As you may have guessed, when only comparing the means from two statistical samples, the Student's *t*-test gives the same results as the ANOVA does. It is when comparing more than two means, the ANOVA test is more reliable.

There are two types of ANOVA tests. The one-way ANOVA is used to determine whether there are any statistically significant differences between the means of three or more independent (unrelated) groups. The two-way ANOVA is used to evaluate data from groups that each has two independent variables. When using statistical software to conduct the ANOVA test, the software will calculate a *p*-value. Based on whether the calculated *p*-value is larger than or less than a preselected α value, the hypothesis of "all the means from the multiple sample groups are statistically equal" can either be rejected or cannot be rejected. The terms of "hypothesis," "reject," "α value," and "*p*-value," are very important statistical concepts that will be discussed in the following sections. Be aware that ANOVA compares all group means and concludes whether there exist differences, but it does not tell which groups are different. There are additional statistical tests such as the Tukey and Sheffe tests that can be used to determine which groups are different, while those statistics are beyond the scope of this book. In our daily work, when we conduct method transfer and need to find which groups produce different means, we can look at the differences between the transferring laboratory and each of the receiving laboratory, by using the Student's *t*-test, or as we will touch later, by simply comparing each of the difference against some preset acceptance criteria.

6.1.5 Hypothesis Setting

The purpose of statistical analysis is to make a statement regarding current or historical events or to make a prediction of future possibilities. Using the statistical language with which by now we are familiar, we can say one big goal for statistical analysis is to make a statement or prediction about a *population* based on the dada of a *statistical sample*. Indeed, observant readers may have already found that the Student's *t*-test is to verify whether there is any statistical *difference* between two means; and the *F*-test is to check whether the two variances are statistically *equal*. The thought process in statistical analysis does not start from collecting data followed by extracting information from the data and then draw conclusions. Rather, the thought process is that the scientists first have some ideas, then they design some experiments to collect data, followed by validating whether the ideas make sense or not by using appropriate statistical models. This thought process, in a statistical language, is called a hypothesis testing. For example, pharmaceutical scientists first thought a drug product can cure a disease, then they formulate a product and design a series of clinical trial studies from which they hope their beliefs can be demonstrated to be correct, i.e. the drug does cure the disease (and with tolerable side effects).

In statistical analyses, there are always two hypotheses, one is called a null hypothesis (H_0), and the other one is called an alternative hypothesis (H_1 or H_a). The alternative hypothesis is sometimes also referred to as a research hypothesis. That is because it is the alternative hypothesis that scientists are trying to prove it is correct, and the null hypothesis is the one that the scientists are trying to reject. The word "null" means to nullify or to disqualify something, or some statement, or some common belief, etc. In statistical language or equations, a null hypothesis always includes an equal sign ($\leq, =, \geq$), and the alternative hypothesis has the other signs ($<, \neq, >$). Depending on the question, or the belief the scientist wants to prove, the testing can be either one direction or two directions, as listed below:

Two tailed (two-sided) hypothesis testing:

Null hypothesis H_0: $\mu_a = \mu_b$
Alternative hypothesis H_1: $\mu_a \neq \mu_b$

One tailed (left-sided) hypothesis testing:

Null hypothesis H_0: $\mu_a \leq \mu_b$
Alternative hypothesis H_1: $\mu_a > \mu_b$

One tailed (right-sided) hypothesis testing:

Null hypothesis H_0: $\mu_a \geq \mu_b$
Alternative hypothesis H_1: $\mu_a < \mu_b$

A hypothesis testing evaluates two mutually exclusive statements regarding a population to determine which statement is best supported by the sample data. Researchers conduct statistical analyses to reject (i.e. nullify, disprove) the null hypothesis. Scientists can either prove their statements or predictions are true when the null hypothesis, i.e. the opposite opinion, can be rejected; or find out that the null hypothesis cannot be rejected and therefore have to declare that the alternative hypothesis cannot be proved. Regardless of what the researcher wants to prove, however, a statistical analysis is always about whether to reject a null hypothesis or whether the statistical analysis fails to reject a null hypothesis. Note that when failing to reject, it does not necessarily mean the null hypothesis is correct, it only means that there is not sufficient evidence to prove it is incorrect. Two types of errors can occur in the conclusions from the hypothesis testing: one is called the <u>*Type I Error*</u>, that is, <u>*reject the null hypothesis when it is true*</u>. The other one is called <u>*Type II Error*</u>, that is, <u>*fail to reject the null hypothesis when it is false*</u>. Type I error is regarded as a false positive error. A false positive error occurs when we regard a difference observed is statistically significant or true, while the difference only occurred by chance. Type II error, on the other hand, is regarded as a false negative error. Not being able to reject a null hypothesis means we cannot accept the alternative hypothesis.

How to set up a hypothesis testing is somewhat confusing for nonstatisticians. The equal (=) sign or unequal (≠) sign does not necessarily have anything to do with whether there is a Yes or No statement in the hypothesis. For example, the hypothesis can be stated as "The two methods are equivalent," in which there is a hidden "Yes" declaration. The same meaning can be expressed as "The two methods are not different," in which there is a "No" message. A third way of saying the same hypothesis is "The difference observed in the two methods is not statistically significant." Some incorrect uses of the statistical tools are due to the wrong expectations set in the hypotheses or improper application of the statistical models. An example is that analytical scientists may try to demonstrate the *equivalency* of the two means by using the Student's *t*-test, while the calculation is designed to verify whether or not the two means are statistically *different*. In contrast, the *F* test is designed to evaluate whether the variances in two groups are *equal*. Remember the one-way ANOVA test is for finding the differences among the means based on the *F* test. The null hypothesis in ANOVA is therefore "all the means are equal." The thought process of setting up the hypothesis only based on what we want to prove can potentially lead to the following confusions:

Step 1: Since we are trying to demonstrate the method is equivalent, we set that as our research hypothesis or alternative hypothesis;
Step 2: Therefore, the null hypothesis is that the method is not equivalent;
Step 3: So, it seems we need to set the null hypothesis as $\mu_a \neq \mu_b$.

But didn't we just learn that the null hypothesis contains only the equal signs? Then let us set the hypothesis in another way:

Step 1: Let us try to demonstrate the method is NOT equivalent, we set that as our research hypothesis or alternative hypothesis;
Step 2: Therefore, the null hypothesis is that the method is equivalent;
Step 3: So, the null hypothesis can be now expressed as $\mu_a = \mu_b$.

Then, we apply the Student's *t*-test to evaluate the equivalency between the two means and forget the *t*-test is for evaluating the difference.

The correct way of setting the null hypothesis is "The difference of the two methods is statistically significant." The equation is expressed as $\mu_a - \mu_b = $ significant, which we will discuss in more details later.

To make it simple for nonstatisticians, we can keep in mind that whether setting as the null or alternative hypothesis for statements such as "Mr. Banana Bread is not guilty," "Mr. Banana Bread is guilty," "The new method is equivalent with the old method," "The new method is not equivalent with the old method," "The treatment by the new drug is effective," or "The treatment by the new drug is not effective," does NOT solely depend on what we want to prove. A piece of useful guidance is that we <u>set up the hypothesis so that the consequence of Type I error is more severe</u>. The following tables show some practical examples to illustrate which statement should be set as the null hypothesis.

Example 6.1   Choose Null Hypothesis in Situations Involving a Court Decision

Option 1:

If we set the null hypothesis H_0 as: **Mr. Banana Bread is innocent (not guilty)**	And the <u>reality</u> is:	
	H_0 is true *Mr. Banana Bread is INDEED innocent (not guilty)*	H_0 is false *Mr. Banana Bread is NOT innocent (guilty)*
What if we reject H_0	Wrong decision (type I error)	Correct decision
What if we fail to reject H_0	Correct decision	Wrong decision (type II error)
Consequence of <u>wrong</u> decision	Innocent Mr. Banana Bread is sent to prison	Guilty Mr. Banana Bread walks away free
Severity of consequence from the <u>wrong</u> decision	**More severe**	Less severe

Option 2:

If we set the null hypothesis H_0 as: **Mr. Banana Bread is not innocent (guilty)**	And the <u>reality</u> is:	
	H_0 is true *Mr. Banana Bread is NOT innocent (guilty)*	H_0 is false *Mr. Banana Bread is INDEED innocent (not guilty)*
What if we reject H_0	Wrong decision (type I error)	Correct decision
What if we fail to reject H_0	Correct decision	Wrong decision (type II error)
Consequence of <u>wrong</u> decision	Guilty Mr. Banana Bread walks away free	Innocent Mr. Banana Bread is sent to prison
Severity of consequence from the <u>wrong</u> decision	Less severe	**More severe**

From nowadays society value system, sending an innocent person to prison is viewed as a more severe consequence. Therefore, the error that leads to that consequence will be set as the Type I error. Rejecting the statement of "Mr. Banana Bread is innocent (not guilty)" when the statement is actually true will cause this error, and thus it is this statement that will be set as the null hypothesis. In other words, the right approach of conducting a trial in courts is: "assuming innocent until proven guilty."

Example 6.2 Choose Null Hypothesis for Method Equivalency Evaluation

Option 1:

If we set the null hypothesis H_0 as: **Method A is not equivalent with method B**	And the *reality* is:	
	H_0 is true	H_0 is false
	Method A is NOT equivalent with method B	*Method A is INDEED equivalent with method B*
What if we reject H_0	Wrong decision (type I error)	Correct decision
What if we fail to reject H_0	Correct decision	Wrong decision (type II error)
Consequence of *wrong* decision	Two different methods are determined as equivalent; contradictory results may be generated by the two methods	Two equivalent methods are determined as inequivalent; the two methods cannot be used interchangeably
Severity of consequence from the *wrong* decision	**More severe**	Less severe

Option 2:

If we set the null hypothesis H_0 as: **Method A is equivalent with method B**	And the *reality* is:	
	H_0 is true	H_0 is false
	Method A is INDEED equivalent with method B	*Method A is NOT equivalent with method B*
What if we reject H_0	Wrong decision (type I error)	Correct decision
What if we fail to reject H_0	Correct decision	Wrong decision (type II error)
Consequence of *wrong* decision	Two equivalent methods are determined as inequivalent; the two methods cannot be used interchangeably	Two different methods are determined as equivalent; contradictory results may be generated by the two methods
Severity of consequence from the *wrong* decision	Less severe	**More severe**

The consequence is more severe when using two inequivalent methods to generate noncomparable data, without knowing the route cause is the choice of methods and potentially declare the product is flawed. The consequence of restricting the freedom of using two interchangeable methods for sample testing is less severe. Therefore, the right null hypothesis is: "Method A is NOT equivalent to Method B." In this method equivalency evaluation case, the right approach is: "assuming guilty until proven innocent." Translate the real-life daily English of "Method A is NOT equivalent to Method B" into a more scientific and statistical English, the null hypothesis is then expressed as "The difference between Method A and Method B is statistically significant."

Example 6.3   Choose Null Hypothesis for New Drug Effectiveness Evaluation

Option 1:

If we set the null hypothesis H_0 as: **The new drug is not effective**	And the *reality* is: H_0 is true *The new drug is NOT effective*	H_0 is false *The new drug is INDEED effective*
What if we reject H_0	We made a wrong decision, which by default is a type I error	Correct decision
What if we fail to reject H_0	Correct decision	We made a wrong decision, which by default is a type II error
Consequence of *wrong* decision	An ineffective drug is regarded as an effective drug and gives patients faked hopes and costs their money and potentially lives for nothing	An effective drug is not able to be approved for patient use; more time or effort may be needed to further study
Severity of consequence from the *wrong* decision	**More severe**	Less severe

Option 2:

If we set the null hypothesis H_0 as: **The new drug is effective**	And the *reality* is: H_0 is true *The new drug is INDEED effective*	H_0 is false *The new drug is NOT effective*
What if we reject H_0	We made a wrong decision, which by default is a type I error	Correct decision
What if we fail to reject H_0	Correct decision	We made a wrong decision, which by default is a type II error
Consequence of *wrong* decision	An effective drug is not going to be approved for patient use; more time or effort may be needed to further study	An ineffective drug is regarded as an effective drug and gives patients faked hopes and costs their money and potentially lives for nothing
Severity of consequence from the *wrong* decision	Less severe	**More severe**

Since the consequence is more severe if we market a nonbeneficial medicine than delay the drug development, the right null hypothesis is: The new drug is NOT effective.

6.1.6 Level of Significance and *p*-Value

As mentioned previously, the Type I error occurs if we reject a null hypothesis when it is actually true. To assess the *probability* of making such an error, we introduce a concept called level of significance, α. The Type II error occurs if we fail to reject a null hypothesis when it is indeed false. The probability of a Type II error is represented as β, which depends on the statistical sample size, level of significance, and the alternative hypothesis. The probability of correctly rejecting a false null hypothesis equals $1 - \beta$ and is called *power*. In more understandable English, *power* is the probability of finding a discrepancy between groups if the difference truly exists. Both α and β reflect a certain level of confidence the scientist has regarding the conclusion inferred from the statistical sample. It tells people how much the scientist considers the conclusion indeed makes sense. Note we do not use the word "correct" or "wrong" here because, in any statistical analysis, all that we do is to conduct a hypothesis testing. There is always a risk that we will make either a Type I or a Type II error. As described in the previous section, the null hypothesis is set based on how we regard the consequence of making either error. If the consequence of making one type of error is considered more severe, we choose that type of error as the Type I error. To give us a measure of how possible we will make the Type I error, or in other words, to tell ourselves how much we can tolerate the consequence of making a Type I error, we come up with this objective measurement, α, the level of significance. An $\alpha = 0.05$ is a commonly accepted number for the willingness to take the risk of making the Type I error. It means that we have a 95% confidence that our conclusion is reflecting the truth, but we are willing to accept that there is a 5% possibility that we are making a mistake. If we regard the consequence of making such a mistake is too severe to accept, we can set the α to be much lower, such as $\alpha = 0.01$. This means we only allow a 1% possibility that we are making a Type I error. An accordingly very heavy-weight aspect regarding any statistical analysis is that, we should be aware that the conclusion from a statistical analysis can be quite different, even flipped, depending on the choice of α. We may decide to reject a null hypothesis (Mr. Banana Bread is innocent) at an $\alpha = 0.1$ (so Mr. Banana Bread will be sent to jail), while find ourselves have a change of mind not to reject the same null hypothesis at an $\alpha = 0.01$ (so Mr. Banana Bread walks away free). Lowering the level of significance does not mean we are error immune. It means although we are lowering the chance of making a Type I error, we are increasing the chance of making a Type II error (Mr. Banana Bread committed a crime but walked away free). Therefore, when designing and planning a study, we have to pre-decide the values of α and β, knowing that inferential statistics is always a balance call between Type I and Type II errors.

Graphically, on a normal distribution curve, the level of significance is represented by the area under the curve that belongs to a so-called *rejection region*.

Figure 6.2 has both sides shaded (two-tailed), which is a nondirectional hypothesis that H_0: no effect, i.e. $\bar{x} = \mu$. If only right side (or left side) is shaded, then it is a directional hypothesis that H_0: sample mean is LESS than the population mean, i.e. $\bar{x} \leq \mu$ (right sided); or H_0: sample mean is LARGER than the population mean, i.e. $\bar{x} \geq \mu$ (left sided). Scientists apply the statistical models such as t-test or F-test to evaluate the data obtained from the statistical sample. If the statistical calculation reveals that the obtained data falls within the *unshaded* region, we conclude that the null hypothesis cannot be rejected. If the statistical calculation reveals that the data falls within the *shaded* region, we reject the null hypothesis. Remember, it is the alternative hypothesis that we are trying to prove. So, making the rejection of the null hypothesis more difficult by allowing smaller levels of significance (such as $\alpha = 0.01$) makes the scientist's job more difficult, and thus the final conclusion is relatively safer. As it is shown on a normal distribution curve, the rejection region is on the tails of the curve, and thus hypothesis analysis is referred to as either a one-tailed or two-tailed testing. On the left side of the curve, $\alpha = 0.05$ means there is a 5% chance that the null hypothesis, $\bar{x} \geq \mu$ is rejected. On the right side, $\alpha = 0.05$ means there is a 5% chance that the null hypothesis, $\bar{x} \leq \mu$, is rejected. If there is no direction, both left and right side of the curve are considered, then there is a total level of significance of $0.05 + 0.05 = 0.1$, i.e. there is a total of 10% chance the null hypothesis of $\bar{x} = \mu$

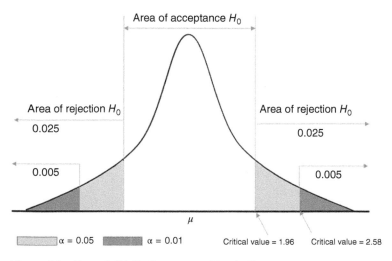

Figure 6.2 Normal distribution curve with rejection areas.

will be rejected. The critical values listed in the figure are for example. A critical value of 2.58 is for $\alpha = 0.01$ for a two-sided hypothesis, and a critical value of 1.96 is for $\alpha = 0.025$ for a two-sided hypothesis. If the statistics (called a z-score) calculated for a two-sided hypothesis which has $\alpha = 0.025$, for example, is 2.00, which is larger than the critical value of 1.96, then the conclusion is that the null hypothesis is rejected at a 95% confidence. On the other hand, since 2.00 is less than the critical value of 2.58 for a two-sided hypothesis which has $\alpha = 0.005$, then the null hypothesis cannot be rejected if the confidence level is set at 99%.

A p-value is also a probability value, but it is not about the probability of making the Type 1 errors like the α value. *p-Values are calculated based on the assumption that the null hypothesis is true.* It is a probability of obtaining a piece of data in the _population_ that is even farther away from the mean than where the α rejection region starts. In Figure 6.2, two α values, 0.05 and 0.01, are shown on both tails of the curve. If the p-value calculated from the test statistics is farther away from the mean, i.e. the p-value falls within the shaded areas represented by $\alpha/2 = 0.025$, then the null hypothesis of $\bar{x} = \mu$ has to be rejected, with a 95% confidence. If the p-value calculated from the test statistics falls within the shaded area represented by $\alpha/2 = 0.005$, then the null hypothesis has to be rejected, with a 99% confidence. On the contrary, if the preset α value is 0.05, and the calculated p-value is 0.030 ($p > \alpha/2$), which means the p-value falls inside the unshaded area represented by $\alpha = 0.05$, i.e., the area of acceptance H_0, then the null hypothesis cannot be rejected. Again, here the important point is that depending on the α value which is supposedly a preselected value, it is possible a p-value is less than one α value, e.g. 0.1, but higher than another α value, e.g. 0.05. Then, the null hypothesis can be rejected if the preselected α value is 0.1, while the null hypothesis cannot be rejected if the preselected α value is 0.05. In some cases, as a simple rule of thumb, a $p < 0.05$ suggests the existence of a statistically significant difference, and a $p > 0.05$ suggests that there is a lack of evidence that there is a statistically significant difference. Detailed calculation of the p-values can be found online or through a literature search.

One caveat is that the p-values are statistical tools and do not necessarily tell the truth about whether there is a true difference or a true significance in analytical data or clinical trial results. p-Values depend upon the sample size. When the sample size is large, results can reach statistical significance (i.e. small p-value) even when the effect is small and analytically irrelevant. On the other hand, with small sample sizes, results can show no statistical significance while the true impact is large and are analytically important. In addition, statistical significance does not take into account the possibility of bias. In the statistical world, a bias should always be corrected before computing the statistical analyses, while in the analytical chemistry world, systematic effects sometimes are not avoidable. But attempts to correct the data by adding back or subtract from the testing results should be carefully scrutinized if not completely forbidden.

In summary, the statement "the null hypothesis will be mistakenly rejected" is, however, very misleading, although you may find that kind of statement in many literature works. It is not "will be," it is actually the scientist "chooses to allow the null hypothesis to be mistakenly rejected." This is an important concept that an analytical scientist should understand before even using any statistical tools. The choice of the level of significance is critical for computing other statistics such as the p-value, confidence interval, prediction interval, and tolerance interval. The conclusions can change dramatically based on the level of significance that the scientist selects.

6.1.7 Confidence Interval, Prediction Interval, and Tolerance Interval

As a scientist working in an analytical development laboratory, practically it is hardly ever we get to know the exact value of the true mean of a population. We only have time to analyze a handful of data and conduct some statistical analyses based on the limited amount of information we get. We need to have some estimation on how much credit we are willing to give to ourselves on the statement we make regarding the true mean. This "credit estimation" is Confidence, which is $(1 - \alpha)$. As we have emphasized multiple times that the level of significance is supposedly a preselected value, the confidence is then the confidence held by the scientist *before* conducting the statistical analysis, not after the scientist has seen the data. However, the choice of a confidence level, i.e. whether the rejection region starts at $\alpha = 0.01$ (1%), 0.05 (5%), or 0.1 (10%), is somewhat arbitrary. The 5% rejection region, or in another word the 95% confidence, is most commonly accepted as a balanced selection to keep both Type I error and Type II error in mind.

Since it is about our credit as a scientist, we do not want to just bet on one value, we preferably set a range for our prediction/conclusion, and hope the true mean residing within that range. This range is called the Interval. As you can see now, the phrase Confidence Interval (CI) is not talking about an interval of a confidence, it is actually talking about two things: (1) a data range (i.e. an interval) that covers the mean (average) of a population; (2) the level of confidence we have regarding the probability that the true mean will fall within that interval. It is not about the interval or range regarding the confidence itself, as many analytical scientists get confused about the phrase. Confidence interval is sometimes called Confidence Limits, in that the upper value called Upper Confidence Limit (UCL) and the lower value called Lower Confidence Limit (LCL). Again, it is not about the limit of confidence. It is about two things: one is the confidence that the scientist has; the other one is the data values that represent the numerical limits of the interval. For example, if the 95% Confidence Interval of the average analytical recovery obtained during a method validation is 92–103%, theoretically you can be 95% confident that the true mean recovery obtained by this method throughout

the method life cycle is within an interval of 92–103%. If we do the method validation again, the actual interval can change based on the new set of validation data,
say, between 90 and 101%. But the true mean recovery possibly is falling within
the overlap of the two intervals. If we keep doing the recovery test and accumulate
a large amount of data, we will find that the interval gets narrower and narrower
since we are getting closer and closer to the true mean recovery, and eventually
the interval reaches to zero.

Now let us see how the confidence interval is calculated. Recall that if σ is the
standard deviation for individual data points, then $\sigma / \sqrt{n}$ is the standard deviation
for averages (also called the standard error of the sample average). As shown in
Figure 6.1, one standard deviation is 68% of the area under the normal distribution
curve, then $\sigma / \sqrt{n}$ means that there is 68% likelihood that the sample mean, $\bar{x}$,
falls within $\sigma / \sqrt{n}$ of the population mean μ. Similarly, there is a 95% likelihood
that the sample mean falls within $2\sigma / \sqrt{n}$ of μ; and a 99.7% likelihood that the
same mean falls within $3\sigma / \sqrt{n}$ of μ. Using statistical language, we say $P(\bar{x}$ is
inside the interval $\bar{x} \pm 2\sigma / \sqrt{n}) = 0.95$. The sentence reads this way: "The probability that $\bar{x}$ is inside the interval $\bar{x} \pm 2\sigma / \sqrt{n}$ is 95%." On the other hand, if we
want to know how much probability the population mean is within an interval
calculated from the sample mean, then we write $P(\mu$ is inside the interval
$\mu = \bar{x} \pm 2\sigma / \sqrt{n}) = 0.95$, which reads "The probability that the population mean, μ,
is inside the interval of the sample mean plus/minus two standard errors is 95%."

However, there is a little caveat in the above probability statement. As mentioned at the beginning of this section, we do not have the luxury to have a large
amount of data to even know the population standard deviation, σ. Therefore,
practically we have to use the statistical sample standard deviation, s_x, and use the
t-distribution (designed for smaller sample size) instead of the normal distribution. The confidence interval is calculated as follows:

$$\bar{x} \pm t \times s_x / \sqrt{n} \tag{6.15}$$

where t is a critical value determined from the t_{n-1} distribution and can be found
from the t table with df $= n - 1$.

Note that a 95% confidence interval does not suggest that the interval you
obtained from the specific set of experimental data that is right in front of
you must have a 95% probability to contain the true mean. The interval calculated from one randomly chosen statistical sample can either contain the true
mean or it does not. A 95% confidence interval means if many statistical
samples are collected, then eventually, statistically about <u>95% of those inter-
vals</u> would contain the true mean. Of course, let us not forget the most relevant
assumption to an analytical scientist, that is, the above interval calculation is
meaningful only if there are no systematic errors (bias) in the data. Imaging if

there is a 10% systematic bias in the assay analyses due to the analytical methodology, then no matter how many statistical samples are tested, the confidence intervals may never contain the true potency of the product. Therefore, some method validation SOPs may mandate that the 100% recovery has to be inside in the calculated confidence interval to ensure that bias is minimized in the method.

Confidence interval is a widely used concept. However, we should keep in mind that a confidence interval is dependent on the sample size, as the Eq. (6.15) shows. The larger n is, the smaller $t \times s_x / \sqrt{n}$ will be, and thus the narrower the interval becomes. In other words, it means that the more data points we have, the closer the obtained sample mean gets to the population mean or the true value. However, all that the confidence interval is telling us is about the mean of the data points. It does not tell us at all the distribution of individual data. For example, if I commute from home to work and I have done that daily for 20 years. I know that the average commuting time is 35 minutes. Since I have a large number of data from those 20 years of daily commuting, I can consider that 35 minutes driving is the true average of the commuting time. In other words, the population mean $\mu = 35$ minutes. So if I take 40 statistical samples from this 20-year-daily-commuting-time database, and make sure the statistical sample size is appropriate for each of those 40 samples, then I can have a 95% confidence, i.e. 95% of those 40 statistical samples from this population (about 38 samples out of the 40 samples) will contain that "magic" commuting time of 35 minutes. However, the intervals of those 40 samples can be all different and the average commuting time of each of those 40 samples can be all different. What the confidence interval focusing on is the conclusion regarding whether or not the population true mean, the 35 minutes, is included in at least 38 out of 40 intervals. Remember, at the time of calculating the confidence interval, we are very likely not aware of the value of the true mean yet.

Opposite to many incorrect considerations or misuse of it, a confidence interval does not have much prediction power since it does not reflect the data scattering in the population. Still using the commuting example, if I have one year of data, and have constructed a confidence interval of 30 ± 15 minutes, I would not be able to say with any confidence whether I do the commuting again for one more year, what will be the chance that my commuting time is within 15–45 minutes. I may obtain a new interval of 40 ± 10 minutes, although the true mean of 35 minutes (we will only know that number after 20 years!) is indeed captured in both this year and next year's intervals.

So, if I want to predict whether I can have some idea about next year's (average) commuting time based on some intervals that I calculated from this year's commuting data, I need some other statistical tools. In other words, I want to have some confidence that the interval I predicted would include the future yet-to-know commuting times for the next year, for the 3rd year, the 4th year, or for the

10th years, etc., although I do not know what exactly the true average time will be (again, to know that, I may have to continue driving for 20 years!). Statistically, it is expressed as: based on the mean and data scattering in a single original study we would like to construct a so-called *prediction interval,* so that we can expect that this interval would capture a certain percentage of the individual data in replication samples. Keep in mind that, similarly, a prediction interval is also talking about two things: one is the probability of the prediction and the other one is the interval within which the future sample data points would fall. What a 95% prediction interval will predict is that if the same statistical sampling is conducted (repeated) the same way in the future for many times, then 95% of those newly observed means from those corresponding statistical samples will fall within an interval that contains the original mean $\bar{x}$. In the commuting example, it means we will be able to say "The prediction shows that 95% of future average commuting times will fall within an interval between $\bar{x} \pm y$ minutes." The $\bar{x}$ is the statistical sample mean obtained from the data on hand, and the y is the interval/range that is predicted. Since a prediction interval accounts for both the sampling error in computing the population mean and data scattering, a prediction interval is always broader than a confidence interval.

The implication of a prediction interval is incredible. This interval tells us if the result from next sets of data obtained from the same batch of products falls within the prediction interval, then the observed difference (if ever) in the original result and the new result is most likely due to the sampling error, which is inevitable in any statistical analysis. On the other hand, if the result falls outside the prediction interval, then we know that in addition to the sampling error, there is too much data scattering, i.e. something else is going on, and investigation is needed. Prediction intervals are often used in regression analysis. The data scattering is predicted to distribute alongside the regression line with a certain interval.

Now we will move one step further. What if not only do we concern about capturing the population mean, but also do we wish to know how possible a future data point falls within a prediction interval, i.e., we want to know how likely a proportion of the population is covered by our prediction? In the commuting example, what if we want to know with certain confidence regarding the chance of completing my future commuting between, say, 35 and 40 minutes, given that I know my commuting time is usually between 25 and 45 minutes? For that purpose, a concept of tolerance interval is introduced. A tolerance interval is a prediction interval. But it does not concern about a single future mean, instead, it considers how likely this interval covers a specified proportion (the 35–40 minutes) of the population (the 25–45 minutes). A tolerance interval is also a confidence interval. But it does not concern about the population mean, it is a confidence interval for a specified proportion of the population. Therefore, there are two variables: one is what percentage of the population the prediction is to cover, and the other one is how much confidence we have in that prediction interval. For example, if we construct a (95%, 90%) tolerance interval, this means with 95% confidence, that at least we have 90% chance to see the future data points fall within the interval we specified. The two variables are independent of each other and can be varied based on desired confidence and coverage. In the commuting example, if we found a (95%, 90%) tolerance interval for the commute is 35–40 minutes, we will say that we are 95% confident that 90% of the future commuting time will be within 35–40 minutes. Note that due to its prediction nature, and due to the lack of crystal balls through which we can see futures, a tolerance interval is usually broad, and is wider than the prediction interval. With the confidence level goes up, either the tolerance interval becomes wider or the probability goes down since an increased ambiguity is needed to keep up with the increased confidence level in order to cover the specified proportion of the population. Therefore, if we want a 99% tolerance interval to cover the 35–40 minutes commuting time, we may have to lower the probability level to, for example, 70%. So we say that we are 99% confident that there is a 70% chance that the future commuting times will fall between 35 and 40 minutes. Or, we can keep the confidence and probability both high (99%, 90%), then we will find the interval becomes wider, such as between 28 and 42 minutes. So it becomes something like this: we know the commuting time is between 25 and 45 minutes and we have 99% confidence that 90% of the future commuting is within 28–42 minutes. By the way, in this example, the tolerance interval of 28–42 minutes is not truly meaningful in real-life scenarios since the range is so broad that it has hardly any predictive value.

Tolerance limits are given by $\bar{x} \pm ks$, with $\bar{x}$ and s denoting the sample mean and the sample standard deviation, respectively, and where k is determined so that one can state with $(1 - \alpha)$% confidence (e.g. 95%) that at least, e.g. 90%, of the data fall within the given limits. The values for k, assuming a normal distribution, have been numerically tabulated. This is commonly stated as something like "a 95% confidence interval for 90% coverage." One application of the tolerance interval in pharmaceutical production is that we can evaluate a manufacturing process by testing the products and compare to the specification. For example, analytical scientists randomly select a sufficient amount of tablets and conduct assay analyses. The mean and standard deviation of the tablet potencies can be calculated. We then request a tolerance interval to be computed with a 95% confidence and covers 99% of the data. The results come back, for example, that the interval meets the confidence level (95%) and the probability level (99%) is between 93 and 107% of the label claim. Then we are 95% confident that 99% of all future tablets manufactured by the current manufacturing process will have potencies that are between 93 and 107%. If the specification is, however, that the potencies must be within 95–105% of the label claim to release the product to the market, then the manufacturing process has to be optimized.

6.2 Application of Statistics – Analytical Method Equivalency

Precision and accuracy are two aspects of any measurement. Precision is regarding the data distribution, and accuracy is regarding the trueness of the measurement. We know by now that precision is represented by variance, standard deviation, and relative standard deviation, etc., and the accuracy is represented by the mean of the sample, standard error, and confidence interval of the mean, etc. When comparing results obtained from a new analytical method with those obtained from the previously established analytical method, both the method precision and accuracy need to be taken into account.

There are different procedures to compare the method equivalency. Some approaches seem to be more or less empirical, but have gained popularity among the industry due to their simplicity [16]. For example, Table 6.2 lists the acceptance criteria for method transfer, intermediate precision, or method comparison studies.

Besides the above simple approach, statistical methods may be desired in some cases. However, as we have already seen that statistical methodology should be used with cautions.

In the hypothesis setting section, we mentioned how to correctly set a null hypothesis. If we set the null hypothesis as $\mu_a = \mu_b$, and want to see whether the two methods are equivalent, we may be tempted to use the Student's t-test. However, using Student's t-test for equivalency determination is rather an inappropriate way of applying this statistical tool. The Student's t-test is designed for finding the existence of statistically significant difference, and is not designed for finding equivalency. Therefore, although the purpose is often to demonstrate the equivalency between two methods, the Student's t-test tries to find the evidence to

Table 6.2 Example empirical acceptance criteria for method comparison.

Methods	Acceptance criteria for **accuracy comparison** (%difference = $100 \times (M_1 - M_2)/$[Average of (M_1, M_2)])
Assay, content uniformity	±2% or ±4%
Dissolution	±2% or ±4%
Impurity, degradation products	±20% or ±25% or ±30%
Methods	Acceptance criteria for **precision comparison** (%RSD)
Assay, content uniformity	≤2% or ≤4%
Dissolution	≤2% or ≤4%
Impurity, degradation products	≤4% or higher on a case by case basis

prove the opposite. In other words, the scientist wants to demonstrate $\mu_a = \mu_b$, while the Student's *t*-test tries to find evidence to against it, by finding the results of the two methods are different. This then defeats the purpose of equivalency evaluation. Moreover, as we have already mentioned many times in this book that in reality the R&D analytical scientists do not have the luxury to obtain a large amount of data to find the true mean of a population. In contrast, usually the scientists only have 6–12 replicate analytical testing results. As many researchers have pointed out, when using the Student's *t*-test, if the statistical sample size is small and the data variation in the results is large, the statistics often does not tell the difference between the two sets of results. See Table 6.3 for example. For convenience the equation used to calculate the *t* value is listed here again.

$$t = \frac{\overline{x} - \overline{y}}{\sqrt{\dfrac{s_x^2}{n_x} + \dfrac{s_y^2}{n_y}}}$$

Table 6.3 contains recovery results from two analysts. Any analytical scientist will see just by looking at the data that the results from Analyst 1 are sporadic (%RSD = 8.23%) and the results from Analyst 2 have both high precision (%RSD = 0.76%) and good accuracy (recovery = 99.4%). If that is a training result from Analyst 1, the conclusion will be that Analyst 1 needs to be retrained. However, the calculated *t* value is a stunning 0.015, much below the 95%

Table 6.3 Example data set for method comparison, *t*-test with 95% confidence.

Analyst 1	Recovery	Analyst 2	Recovery
Sample 1	90.0	Sample a	99.2
Sample 2	105.1	Sample b	100.1
Sample 3	95.2	Sample c	99.9
Sample 4	109.4	Sample d	98.2
Sample 5	91.5	Sample e	99.9
Sample 6	105.2	Sample f	98.8
Average	99.4	Average	99.4
Standard deviation	8.18	Standard deviation	0.75
%RSD	8.23%	%RSD	0.76%
T-Table critical value (df = $10 t_{0.975}$)	2.228	Conclusion from *t*-test	Two sets of recovery results *are equivalent*
Calculated *t*	0.015		

confidence critical value of 2.228 for a group of data that has a freedom of 10 ($n_1 + n_2 - 2 = 6 + 6 - 2 = 10$). The conclusion from the Student's t-test is that Analyst 1 has successfully passed the training. If the recovery results are from a method comparison exercise, then the conclusion from the statistics is that the two methods are equivalent.

On the other hand, if there are lots of data at hand and the variation in the results is small, which means potentially two very precise results or methods are being compared, the Student's t-test tends to find that a difference does exist. The trouble is although the results show a statistically significant difference, it is analytically irrelevant. But we may still have to do a lot of explanation why we think it is not a deviation or out of specification. See Table 6.4 for example.

Table 6.4 contains recovery results from the same two analysts. This time the results from both Analyst 1 and Analyst 2 are precise and accurate. If that is a training result from Analyst 1, the conclusion will be that Analyst 1 is able to deliver the same quality results as Analyst 2 does. However, the calculated t value is a little higher than the critical value. The conclusion from the Student's t-test is that Analyst 1 does not pass the training! If the recovery results are from a method comparison exercise, then the conclusion from the statistics is that the two methods are NOT equivalent.

The above examples emphasize again the importance of not conducting any statistical analysis in a "black box." Assistance from professional statisticians is desired.

If we follow the approach of setting up the null hypothesis described in this book, we know the correct null hypothesis for method equivalency evaluation is "Method

Table 6.4 Example data set for method comparison, t-test with 95% confidence.

Analyst 1	Recovery	Analyst 2	Recovery
Sample 1	100.0	Sample a	99.2
Sample 2	99.9	Sample b	100.1
Sample 3	100.1	Sample c	99.9
Sample 4	100.2	Sample d	98.2
Sample 5	100.1	Sample e	99.9
Sample 6	100.0	Sample f	98.8
Average	100.1	Average	99.4
Standard deviation	0.10	Standard deviation	0.75
%RSD	0.10%	%RSD	0.76%
T-table critical value (df $= 10 t_{0.975}$)	2.228	Conclusion from t-test	Two sets of recovery results are *NOT equivalent*
Calculated t	2.463		

A is NOT equivalent with Method B," or "Method A is different from Method B." But in the meantime, we also know that we cannot set the null hypothesis as $\mu_a \neq \mu_b$. Some readers may have already figured out that since the other way of setting the null hypothesis is $\mu_a \geq \mu_b$ or $\mu_a \leq \mu_b$, then if we turn our statement into a more statistic sound language, such as, "the difference between Method A and Method B is statistically significant," then it leads us to set the null hypothesis as below:

$$\left|\mu_a - \mu_b\right| \geq \theta$$

or

$$\mu_a - \mu_b \leq \theta_L \text{ and } \mu_a - \mu_b \geq \theta_U$$

where θ_L and θ_U are predefined as the lower and upper acceptable difference limits for equivalence, respectively. When θ_L and θ_U are the same in number with different signs, they are symmetrical limits for the equivalence test. For example, if $\theta_L = -2\%$ and $\theta_U = 2\%$, the limits are either expressed as

$$\left|\mu_a - \mu_b\right| \geq 2\%$$

or

$$\mu_a - \mu_b \leq \theta_L = -2\% \text{ and } \mu_a - \mu_b \geq \theta_U = 2\%$$

Therefore, the task becomes whether we can reject the null hypothesis that the difference between the methods is as large or larger than some pre-set boundaries, θ. To do that, we can use the Student's t-test, but instead of a one-sided (one-tailed) or two sided (two-tailed) testing, here it is a *Two One-Sided t tests*, that is, one side is $\geq \theta_U$ and the another one side is $\leq \theta_L$. The alternative hypothesis is then expressed as

$$H_a : \theta_L < \mu_a - \mu_b < \theta_U$$

This statistical treatment is well known in the pharmaceutical industry as Schuirmann's two one-sided test, which has been historically used to assess bioequivalence data.

Another way of assess the difference is to use the confidence interval approach as shown below:

$$\text{CI} = \left(\bar{x} - \bar{y}\right) \pm t_{0.10\left(n_x + n_y - 2\right)} S_p \sqrt{\frac{1}{n_x} + \frac{1}{n_y}} \tag{6.16}$$

In which $\left(\bar{x} - \bar{y}\right)$ is the calculated difference in the sample means; $t_{0.10\left(n_x + n_y - 2\right)}$ is the t value for 95% confidence for $n_x + n_y - 2$ degree of freedom (note that for 95% confidence, the α is 0.10 for a two one-sided test); n_x is the sample size for the first data set; n_y is the sample size for the second data set; and s_p is the pooled standard deviation.

The calculations of the upper and lower limits of the confidence intervals can be found in the appendix of USP general chapter <1010> "Analytical Data – Interpretation and Treatment." Example calculations for comparison of method duct a 5% one-sided test). The calculation first obtains a ratio between the variances of the two methods:

$$\text{Ratio} = \left(\text{alternative method variance}\right) / \left(\text{current method variance}\right) \qquad (6.17)$$

Then the lower limit of the confidence interval is calculated as below:

$$\text{LCL} = \text{ratio} / F_{0.05} \qquad (6.18)$$

and the upper limit of the confidence interval is calculated as below:

$$\text{UCL} = \text{ratio} / F_{0.95} \qquad (6.19)$$

If the confidence interval calculated by Eq. (6.16) is within the acceptance limits calculated from Eqs. (6.18) to (6.19), the null hypothesis of "The difference between Method A and Method B is statistically significant" is then rejected, and the two methods are considered equivalent. As some researcher pointed out, because the magnitude of the calculated confidence interval increases as the pooled standard deviation increases, Schuirmann's two one-sided test is more likely to conclude that there is a lack of equivalence in the presence of too much variation, the opposite of the effect seen with the two sample Student t-test.

For convenience and information purpose, some F values are listed here. In the USP <1010>, the example is a method comparison based on 20 independent runs per method. The $F_{0.05} = 2.168$ and $F_{0.95} = 0.461$. For a method comparison based on 6 independent samples, the $F_{0.05} = 5.050$ and $F_{0.95} = 0.198$. For a method comparison based on 12 independent samples, which is a common sample size in pharmaceutical industry for method equivalency evaluation, the $F_{0.05} = 2.818$ and $F_{0.95} = 0.3543$.

To evaluate whether the limits set for the UCL and LCL are reasonable, people can use tolerance limits. The tolerance interval can be calculated by the following formula: $\bar{x} \pm ks$. The example provided in the USP <1010> describes a situation that if the population mean and the standard deviation are both unknown, but a statistical sample of 50 data points that produced a mean and standard deviation of 99.5 and 2.0, respectively. These values were calculated using the last 50 results generated by this specific method for a particular (control) sample. In that example, the value of K required to enclose 95% of the population with 95% confidence for 50 samples is 2.3828. The tolerance limits are calculated as follows:

99.5 ± 2.382 × 2.0; hence, the tolerance interval is (94.7, 104.3). The next step is to compare the obtained tolerance interval against the specification limits to find out the smallest confidence interval that is allowed. In the example, assume the specification interval for this method is (90.0, 110.0), and the following quantities can be defined: the lower specification limit (LSL) is 90.0, the upper specification limit (USL) is 110.0, the lower tolerance limit (LTL) is 94.7, and the upper tolerance limit (UTL) is 104.3. Calculate the acceptable difference, (δ), in the following manner:

$$A = LTL - LSL \text{ for } LTL \geq LSL \left(A = 94.7 - 90.0 = 4.7\right);$$

$$B = USL - UTL \text{ for } USL \geq UTL \left(B = 110.0 - 104.3 = 5.7\right);$$

$$\text{And } \delta = \text{minimum}\left(A, B\right) = 4.7.$$

With this choice of δ, and assuming the two methods have comparable precision, the confidence interval for the difference in means between the two methods (alternative-current) should fall within −4.7 and +4.7 to claim that no important difference exists between the two methods.

On a second note regarding the preset boundaries, however, they are the limits set by the scientists based on practical ranges within which the method equivalency is considered acceptable. Interestingly, when thinking in that way, it comes back to Table 6.2. Indeed, one important take-home message is that as an analytical scientist, our job is to develop robust, precise, and accurate methods. If the methods are scientifically trustworthy and practically user-friendly, the use of statistical analysis in date treatment is, somewhat ironically, not an absolute necessity.

In addition, extra cautions are needed when using statistical approach to evaluate the equivalency in analyses of degradation products or impurities where analytical errors are inevitably large.

6.3 Application of Statistics – Stability Data Trending

ICH Q1E EVALUATION FOR STABILITY DATA provides guidelines regarding how to perform stability data trending evaluation. The basis of the evaluation is that, in general, quantitative changes in some chemical attributes (e.g. assay, degradation products, preservative content) of a drug substance or product can be assumed to follow zero-order kinetics during long-term storage. Statistical approaches such as linear regression and poolability testing can be applied. Figure 6.3 shows an example diagram similar to the one provided in the guideline. The regression line is for assay of a drug product with upper and lower acceptance criteria of 110% and 90% of label claim, respectively, with 12 months of long-term data and a proposed shelf life of 24 months. In this example, two-sided 95% confidence limits for the mean are applied

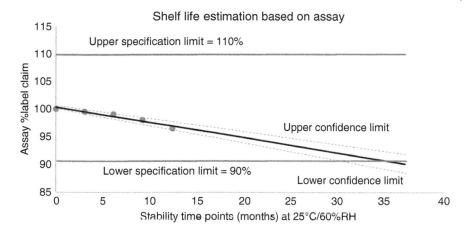

Figure 6.3 Shelf life estimation based on assay analysis with 95% confidence level.

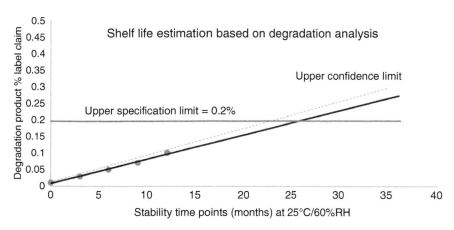

Figure 6.4 Shelf life estimation based on degradation product analysis with 95% confidence level.

because it is not known ahead of time whether the assay would increase (e.g. in the case of an aqueous-based product packaged in a semi-permeable container) or decrease with time. The lower confidence limit intersects the lower acceptance criterion at around 30 months, while the upper confidence limit does not intersect with the upper acceptance criterion. Therefore, the proposed shelf life of 24 months can be supported by the statistical analysis of the assay values.

When data for an attribute with only an upper or a lower acceptance criterion are analyzed, the corresponding one-sided 95% confidence limit for the mean is recommended. Figure 6.4 shows the regression line for a degradation product in a drug product with 12 months of long-term data and a proposed shelf life of 24 months, where the acceptance criterion is not more than 0.2%. The upper one-sided 95% confidence limit for the mean intersects the acceptance criterion at slight shorter than 24 months. Therefore, the proposed shelf life of 24 months cannot be supported by statistical analysis of the degradation product data.

References

1 U.S. Food and Drug Administration (FDA) (July 1988). Guidance for Industry Format and Content of the Clinical and Statistical Sections of an Application.

2 U.S. Food and Drug Administration (FDA) (September 1998). Guidance for Industry E9 Statistical Principles for Clinical Trials.

3 U.S. Food and Drug Administration (FDA) (February 2001). Guidance for Industry Statistical Approaches to Establishing Bioequivalence.

4 U.S. Food and Drug Administration (FDA) (March 2007). Guidance for Industry and FDA Staff Statistical Guidance on Reporting Results from Studies Evaluating Diagnostic Tests.

5 U.S. Food and Drug Administration (FDA) (February 2010). Guidance for Industry Guidance for the Use of Bayesian Statistics in Medical Device Clinical Trials.

6 Analytical Lifecycle: USP <1210> "Statistical Tools", United States Pharmacopeia (USP) General Chapter <1210> Statistical Tools for Analytical Procedure Validation. The United States Pharmacopeial Convention. Official 1 May 2018.

7 Pramod, K., Tahir, M.A., Charoo, N.A. et al. (2016). Pharmaceutical product development: a quality by design approach. *International Journal of Pharmaceutical Investigation* 6 (3): 129–138.

8 Politis, S.N., Colombo, P., Colombo, G., and Rekkas, D.M. (2017). Design of experiments (DoE) in pharmaceutical development. *Drug Development and Industrial Pharmacy* 43 (6): 889–901.

9 Belouafa, S., Habti, F., Benhar, S. et al. (2017). Statistical tools and approaches to validate analytical methods: methodology and practical examples. *International Journal of Metrology and Quality Engineering* 8 (9) https://doi.org/10.1051/ijmqe/2016030.

10 Muth, J.E.D. (2014). *Basic Statistics and Pharmaceutical Statistical Applications*, CRC Press Pharmacy Education Series, 3e. CRC Press.

11 Walfish, S. (2006). Analytical methods: a statistical perspective on the ICH Q2A and Q2B guidelines for validation of analytical methods. *BioPharm International* 19 (12): 28–36.

12 United States Pharmacopeia (USP) General Chapter <32> General Notices and Requirements Applying to Standards, Tests, Assays, and Other Specifications of the United States Pharmacopeia. The United States Pharmacopeial Convention. Official 1 May 2012.

13 Dahiru, T. (2008). P – value, a true test of statistical significance? A cautionary note. *Annals of Ibadan Postgraduate Medicine* 6 (1): 21–26.

14 Amrhein, V. and Greenland, S. (2018). Remove, rather than redefine, statistical significance. *Nature Human Behaviour* 2: 4.

15 Nagele, P. (2003). Misuse of standard error of the mean (SEM) when reporting variability of a sample. A critical evaluation of four anaesthesia journals. *British Journal of Anaesthesia* 90 (4): 514–516.

16 Chambers, D., Kelly, G., Limentani, G. et al. (2005). Analytical method equivalency an acceptable analytical practice. *Pharmaceutical Technology* 29: 64–80.

7

Thoughts on Conventional Chromatography Practices

The purpose of one analytical method is to continuously provide accurate and precise analytical results for a specific product. The method life cycle management starts with method development, followed by method validation and transfer, and then moves forward to routine analyses. During the regular use of the method, modification or improvement may become necessary. The method completes its mission and may retire when the product life cycle ends.

When considering an HPLC method, we need to keep in mind that it contains two perspectives: (1) chromatographic procedures and (2) sample preparation. Both aspects contain parameters or attributes that have hidden assumptions that sometimes may not be very obvious to the analytical scientists and can produce inaccurate results if those assumptions do not hold. This chapter will briefly discuss some of the general method parameters and the critical assumptions behind conventional practices.

7.1 Linear Regression

ICH Q2 states that the accuracy of an analytical procedure expresses the closeness of agreement between the value found and the value which is accepted either as a conventional true value or an accepted reference value. The accuracy should be established across the specified range of the analytical procedure. Typically, the analytical range should cover 70–130% of the label claim in order to analyze the drug product content uniformity; 80–120% of the label claim if only the potency of a drug product or a drug substance is concerned, and 80–120% (or wider, depending on the necessity of obtaining release profiles) for dissolution testing. For an impurity/degradation analysis, the analytical ranges should at least cover from the reporting level of the impurity to 120% of the specification. Broader

Analytical Scientists in Pharmaceutical Product Development: Task Management and Practical Knowledge, First Edition. Kangping Xiao.
© 2021 John Wiley & Sons, Inc. Published 2021 by John Wiley & Sons, Inc.

ranges, especially for impurity/degradation product analyses, should be attempted in case a need for increasing the specification arises.

Although some methods use internal standards, the majority of analyses quantitate the analyte in the sample solutions against the response of an external reference standard. A calibration curve of the external reference standard is established and used for quantitation of the analyte in sample. Regardless of the source of the signal, a calibration curve that an analytical scientist in a pharmaceutical product development team encounters in daily work is most likely a straight line. The values on the y-axis usually represent the detector responses, and the numbers on the x-axis represent the analyte amounts/concentrations in sample solutions. On this x–y 2D plot, the analyte amounts/concentrations are so-called *Independent Variables*, while the detector responses are *Dependent Variables*. The detector signals are dependent on the amounts/concentrations of the analyte.

Linear regression analysis is a statistical approach to assess the relationship between the independent variables and the dependent variables. The most simple equation that describes a linear regression is $y = ax + b$, where a = slope of the straight line, also known as regression coefficient; $b = y$-intercept, which is a y value through which the linear line crosses. The y-intercept is the response from the detector when there is no target analyte in the sample solutions. The y-intercept can be positive and negative, both of which indicate some matrix effects from the sample solutions. The mathematical method widely used for linear regression is the Linear Least Squares Regression. Without going into its modeling details, a least-squares regression is first to square the distance between each found value of the individual dependent variables to the predicted value on the regression line, and then finds out the smallest sum of the thus obtained squares. Figure 7.1 illustrates a linear regression for a group of data points that spread on both sides of a linearity curve.

The plot shown in Figure 7.1 is called a scatter plot. In regression analysis, the distance between the observed value of the dependent variable (y) and the predicted value ($\hat{y}$) is called the Residual. Although visual exam of the linear regression fitting is a viable approach and is suggested even in such as ICH guidelines, a residual plot can provide more graphic effect. The residual plot is a graph that shows the residuals on the vertical axis and the independent variable on the horizontal axis. If the points in a residual plot disperse randomly around the horizontal axis, a linear regression model is appropriate for the data; otherwise, a nonlinear model is more appropriate. The residual plot below (Figure 7.2) is constructed based on the above scatter plot, and you can see there is no specific pattern in the data scattering. Therefore, it is appropriate to use the linear regression for data fitting.

The linear regression imposes a linear relationship between the analyte amounts/concentrations and the detector responses, while it does not indicate

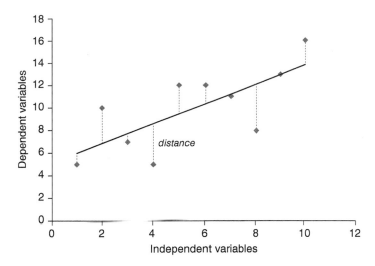

Figure 7.1 Illustration of a linear regression between dependent variables and independent variables.

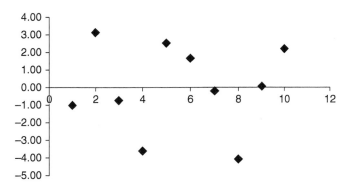

Figure 7.2 A residual plot with a random distribution of data points.

how linear the relationship is. To quantitate the linearity, a *correlation coefficient*, r, is introduced to represent how strong a relationship is between those two variables. Pearson Product Moment Correlation (PPMC) equation calculates the strength of a linear relationship:

$$r = \frac{\sum_i (x_i - \overline{x})(y_i - \overline{y})}{\sqrt{\sum_i (x_i - \overline{x})^2} \sqrt{\sum_i (y_i - \overline{y})^2}} \tag{7.1}$$

The value of r is between -1 to $+1$, with -1 representing a perfect negative, and $+1$ representing a perfect positive linear relationship. When $r > 0.90$, it is generally regarded that a genuine positive linear relationship exists between the two variables. However, as will be seen in the following section, for pharmaceutical assay analysis, $r > 0.999$ may be required.

The *coefficient of determination*, r^2, is used to evaluate the percentage of the data points that can be accurately described by the linear regression equation. If $y = ax + b$, then the r^2 is calculated in the following equation:

$$r^2 = \frac{a^2\left[\sum x^2 - \dfrac{(\sum x)^2}{n}\right]}{\sum y^2 - \dfrac{(\sum y)^2}{n}} \tag{7.2}$$

In practice, many analytical laboratories quite often mistakenly used the r^2 interchangeably with the r. The coefficient of determination is useful because it provides information about how much the dependent variables, i.e. the detector responses, change with the independent variables, i.e. the amounts/concentrations of the analyte in the sample solutions. If r^2 is 0.995, it means that there is 99.5% of the total variation on the y-axis is explainable by the linear regression model. It just so happens that the mathematical value of the *coefficient of determination* is the square of the *correlation coefficient*.

7.2 Response Factor, Linearity Slope, and y-Intercept

A response factor is a ratio between the response of the detector over the analyte amount or concentration.

$$\text{Response factor} = \frac{\text{response}}{\text{amount or concentration}} = \frac{y}{x} \tag{7.3}$$

In practice, the slope of a linearity curve often represents the response factor (RF). However, the linear regression equation $y = ax + b$ clearly indicates that y/x does not equal to a, the slope. Mathematically, the only time when the slope equals y/x is when b, the intercept, is zero. Let us see a simple example of a linear curve of $y = 3x + 4$ (Figure 7.3). In Table 7.1, the columns on the left side contain the values of the x-axis and the y-axis. The slope is 3 for all three curves,

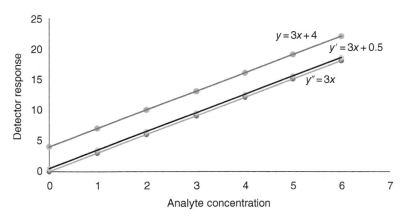

Figure 7.3 Hypothetic linearity plots between detector responses (y) and analyte concentrations (x), with the same linearity slope and different y-intercepts.

and the linear regression is a perfect $r = 1.00$. The RFs, which are represented by the values of y/x, y'/x, and y''/x, for the three curves in Figure 7.3, respectively, are listed in the columns that are right to the columns of the responses of y, y', and y''. Now let us see how much impact a y-intercept can have on method accuracy.

It is apparent in Table 7.1 that the values of RF_1 are drastically different at different x levels, i.e. the RFs vary considerably at different analyte concentrations. The discrepancy between RF_1 and the slope ($a = 3$) can be more than 200%. Mathematically the existence of a y-intercept of 4 does not allow us to use the slope to represent the RFs. From the analytical science point of view, it means that at zero analyte concentration, there exists a response of 4, from the sample matrix. Since, in reality, it is rare for a linearity curve that is plotted from a set of experimental data to really go through zero, a certain value of y-intercept is inevitable. Therefore, we can think that if we have a reasonably small y-intercept, such as $b = 0.5$ in the above example, we will have a new linear relationship of $y' = 3x + 0.5$ (Figure 7.3). The new RFs, i.e., the column of RF_2 (y'/x), are listed in Table 7.1. As you can see that the resulted discrepancy between y'/x and the slope, a, is smaller. The examples numbers in Table 7.1 also reveal that the y-intercept creates a severer discrepancy between the RFs (both RF_1 and RF_2) and the slope at lower analyte concentrations. Obviously, RF_3 is the same as the slope since the y-intercept is zero, as shown in the table (y''/x). Therefore, it is important to list a y-intercept acceptance criterion as one of the critical parameters in method validation.

Table 7.1 Discrepancy between response factors and linearity slopes.

x-Axis	y-Axis	y′-Axis	y″-Axis	$RF_1 = y/x$	$RF_2 = (y')/x$	$RF_3 = (y'')/x$	% [RF_1/slope (a = 3)]	% [RF_2/slope (a = 3)]	% [RF_3/slope (a = 3)]
0	4	0.5	0						
1	7	3.5	3	7.00	3.50	3.00	233.3	116.7	100.0
2	10	6.5	6	5.00	3.25	3.00	166.7	108.3	100.0
3	13	9.5	9	4.33	3.17	3.00	144.4	105.6	100.0
4	16	12.5	12	4.00	3.13	3.00	133.3	104.2	100.0
5	19	15.5	15	3.80	3.10	3.00	126.7	103.3	100.0
6	22	18.5	18	3.67	3.08	3.00	122.2	102.8	100.0

7.3   Relative Response Factor, Linearity Slope, and *y*-Intercept

Ideally, when measuring the concentration of an analyte, we would like to compare the analyte responses in the sample and the working standard solutions. However, when analyzing impurities or degradation products, we find either the reference standards are scarcely available, or they are quite expensive. One approach to cope with that situation is that, instead of using the impurity reference standard every time, we can use Relative Response Factors (RRFs) in routine analysis. An RRF is a ratio of RFs between an impurity/degradation product and the active pharmaceutical ingredient (API; or a different molecule, if appropriate). Theoretically, for RRF estimation, the RFs of both impurity/degradation product and the API should be determined by spiking the compounds in the sample matrix (i.e. placebo). In practice, although the RF of impurity/degradation product is usually determined by spiking the compound into the sample matrix, the RF of the API working standard can be determined in diluent without adding a placebo. The reason is simply due to the potential unavailability of the placebo at the quality control laboratories. Therefore, during routine analysis, the API working standard solutions are usually prepared without a placebo. Since a calculated RRF must cover a wide concentration range, we tend to use the slope of an impurity linearity curve to represent the RF of that impurity analyte, and the slope of the API linearity curve to represent the RF of the API. The RRF determination of an impurity/degradation product can be part of the linearity study during the method validation. The solution concentration range of the impurity/degradation product usually covers from the limit of quantitation (LOQ) to 120% or even 200–300% of the specification limit. The much broader impurity linearity range is to cover future potential increases in the specification limit. For example, if the impurity/degradation product specification limit is 0.5%, the linearity range determined during the method validation can be from LOQ level to 1.0% (200% of 0.5%) or 1.5% (300% of 0.5%), to cover potential increase in the specification limit, up to 300%, without having to validate the method again in the future. The linearity range of the API working standard should be comparable to the concentrations of the impurity/degradation product in the sample solution, but can also extend to cover a much broader range. For example, although the level of the impurity/degradation product ranges from 0.05 to 0.5%, the linearity range of the API working standard can be from 0.05 to 10%. The high concentration level, such as 10%, can reduce the risk of failing the working standard injection precision requirement of the system suitability during routine analysis.

One critical aspect, as we just discussed, is that the *y*-intercept has to be sufficiently small to enable us to use linearity slopes to represent RFs. That is also true for obtaining accurate RRF values.

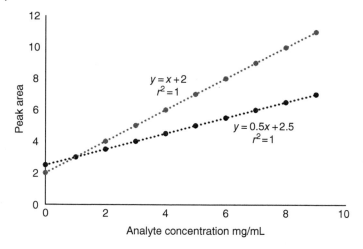

Figure 7.4 Hypothetical linearity plots for relative response factor determination, with large *y*-intercepts.

Figure 7.4 contains two linearity plots for RRF determination, with large *y*-intercepts. From the visual examination, we already know that the RF ratios across the entire concentrations are different since the two linearity curves intersect each other. In Table 7.2 some example peak area and analyte concentration values calculated from the equations listed in the figure, clearly show how much discrepancy can exist between the RRF determined by slope and the RRFs calculated at each concentration level, when the *y*-intercepts are not small.

Table 7.2 Discrepancy between RRFs obtained from ratios of response factors and the ratio of linearity slopes.

Concentration (mg/mL)	Peak area 1	Peak area 2	Response factor 1	Response factor 2	RRF obtained by slope ratio	RRF obtained by RF$_2$/RF$_1$
0	2	2.5	—	—	—	
1	3	3	3	3	0.5	1
2	4	3.5	2	1.8	0.5	0.88
3	5	4	1.7	1.3	0.5	0.8
4	6	4.5	1.5	1.1	0.5	0.75
5	7	5	1.4	1	0.5	0.71
6	8	5.5	1.3	0.9	0.5	0.69
7	9	6	1.3	0.9	0.5	0.67
8	10	6.5	1.3	0.8	0.5	0.65
9	11	7	1.2	0.8	0.5	0.64

As you may have already known by now that if the two linearity curves both have zero *y*-intercept, then the RRFs obtained from the ratio of slopes and by calculating at each concentration levels are exactly the same (Figure 7.5 and Table 7.3).

Although mathematically, the above discussions are quite simple and straightforward, in practice, the importance of obtaining a linearity curve with a small *y*-intercept could be overlooked. To minimize the bias of the method, analytical scientists have to reduce the interference from the sample matrix, to obtain smaller *y*-intercepts.

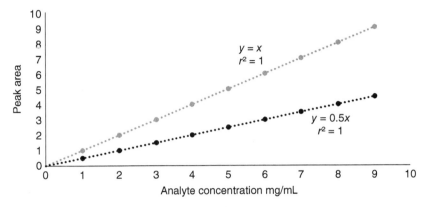

Figure 7.5 Hypothetical linearity plots for relative response factor determination, with zero *y*-intercepts.

Table 7.3 No discrepancy exists between RRFs obtained from response factor ratios and the linearity slope ratio, when the *y*-intercept is zero.

Concentration (mg/mL)	Peak area 1	Peak area 2	Response factor 1	Response factor 2	RRF obtained by slope ratio	RRF obtained by RF_2/RF_1
0	0	0	—	—	—	—
1	1	0.5	1	0.5	0.5	0.5
2	2	1	1	0.5	0.5	0.5
3	3	1.5	1	0.5	0.5	0.5
4	4	2	1	0.5	0.5	0.5
5	5	2.5	1	0.5	0.5	0.5
6	6	3	1	0.5	0.5	0.5
7	7	3.5	1	0.5	0.5	0.5
8	8	4	1	0.5	0.5	0.5
9	9	4.5	1	0.5	0.5	0.5

7.4 Linearity and Method Accuracy

Method accuracy for analysis of a target analyte can be evaluated/represented by the recovery results of this compound. A recovery is a ratio between "how much measured" against "how much prepared." The "how much prepared" represents a theoretical value (amount or concentration) of the target analyte that an analytical scientist prepared in a solution for analysis. In a method validation set up, the solutions are either prepared by using a real finished product, or by spiking a predefined amount of the target analyte into a placebo solution. The "how much measured" is the analytical result of the target analyte in the prepared solution. As discussed previously, the quantitation is by comparing to the RF of an external reference standard of the same kind or by comparing to the RF of an external reference standard of a different molecule but corrected with a correction factor (relative response factor).

In the above few sections, we have seen the importance of having a small y-interception. Now let us ask another question. What if the linearity is not perfect? Or, how linear is linear?

Figure 7.6 shows a linearity curve that has a good linearity reflected by a decent $r = 0.99$ ($r^2 = 0.98$). The absolute value of the y-intercept is 0.0108, which is much less than 2% of the largest response of 1.3. As a common practice, when performing assay analysis, a y-intercept of less than 2% of the analyte response at 100% analytical concentration is considered acceptable. It seems reasonable to accept this linearity curve to perform quantitation, i.e. to obtain the measured concentrations. The linearity is decent, and the y-intercept is small. However, the calculated concentrations are sporadic, as shown by the percent recoveries in Table 7.4.

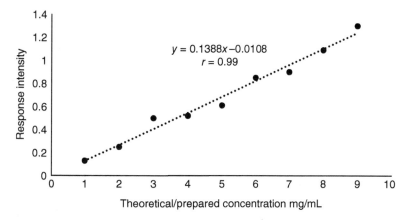

Figure 7.6 Hypothetical linearity plot with a correlation coefficient of 0.99 and a y-intercept of 0.0108.

Table 7.4 Recoveries calculated based on the linearity curve in Figure 7.6.

True concentration (mg/mL)	Detector response	Measured concentration (mg/mL)	Recovery (%)
1	0.13	1.01	101
2	0.25	1.88	94
3	0.5	3.68	123
4	0.52	3.82	96
5	0.61	4.47	89
6	0.85	6.20	103
7	0.9	6.56	94
8	1.09	7.93	99
9	1.3	9.44	105

Table 7.5 Recoveries calculated based on the linearity curve in Figure 7.7.

True concentration (mg/mL)	Detector response	Measured concentration (mg/mL)	Recovery (%)
1	0.134	0.98	98
2	0.25	1.96	98
3	0.37	2.98	99
4	0.5	4.09	102
5	0.61	5.02	100
6	0.73	6.04	101
7	0.84	6.98	100
8	0.95	7.91	99
9	1.08	9.02	100

When conducting assay/potency analysis, the acceptance criterion for the method accuracy is usually within ±2%. The quantitation, based on the calibration curve in Figure 7.6, does not meet that requirement. The results may be acceptable for impurity/degradation product analysis, where the method accuracy acceptance criterion is much broader. Some companies have a wide range between 70 and 130%, and some other companies may accept recoveries between 75 and 125%, for impurity analysis.

The method accuracy improves dramatically (Table 7.5) with a correlation coefficient of 0.9998 (Figure 7.7). The recoveries in Table 7.5 are within ±2% across a

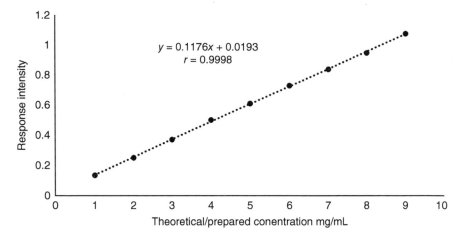

Figure 7.7 Hypothetical linearity plot with a regression coefficient of 0.9998.

broad concentration range. Therefore, to accurately assay the potency of an API, the detector response must be sufficiently linear with a correlation coefficient $r > 0.999$. In practice, if a method has to cover a rather broad analytical concentration range, it is not necessary to measure all the recoveries using one linearity curve. The range can be split into narrower ranges, each with its own linearity curve for quantitation. In addition, the solution preparations at different concentration levels may be modified to minimize matrix effects, or to increase signal-to-noise ratios, etc.

7.5 Injection Precision in System Suitability

Different companies may have different system suitability requirements defined in the SOPs. Injection precision of the working and bracketing standards is one of the most common system suitability parameters. The working standard injections refer to the first five or six standard injections before the sample injections. It seems there is not much difference in the requirement for those first five or six working standard injections that the %RSD of the peak areas usually must be within 2.0%. Difference in practice exists when evaluating the injection precision of the overall sample sequence. The bracketing standard injections refer to the ones that are made in-between sample injections and at the end of the sample sequence. There are around three mainstream approaches for injection precision evaluation. The first one is that the overall %RSD must be no more than 2.0% calculated using all the working and bracketing standard injections throughout the entire sample sequence. The second approach demands a no more than 2.0% difference between the response of each bracketing standard and the average response of the first five or six working standard injections. The third practice is that the %difference in responses has to be no more than 2.0% for each pair of the bracketing standard injections. Although each practice has its logic and should be used on a case-by-case basis, the second practice is a balanced approach. Table 7.6 presents some numbers for illustration purposes.

As the data in Table 7.6 indicate, the risk of using the overall %RSD approach is the potential of overlooking the unacceptable variations in the bracketing standard injections. The peak areas of the first five injections are close to each other that produce an excellent %RSD of 0.11%. Then, the injection precision starts to deteriorate, beginning with the sixth injection. "Luckily," however, the overall injection precision shows a %RSD of 1.99%, which is just less than the 2.0% acceptance criterion. In this case, however, any experienced analytical scientist knows

Table 7.6 Example peak areas of and %differences between working and bracketing standard Injections.

Standard injection	Peak area	%Difference[a] against average working standard	%Difference[a] between each pair of bracketing standard
Injection 1	1000	—	—
Injection 2	1001	—	—
Injection 3	1002	—	—
Injection 4	1000	—	—
Injection 5	999	—	—
Average working standard	*1000.40*	—	—
%RSD	*0.11%*	—	—
Injection 6	970	*−3.09%*	—
Injection 7	1035	*3.40%*	*6.48%*
Injection 8	1020	*1.94%*	*−1.46%*
Injection 9	975	*−2.57%*	*−4.51%*
Overall average	*1000.22*	—	—
Overall %RSD	*1.99%*	—	—
Acceptance criterion	**%RSD ≤2.0%**	**%Difference ≤2.0%**	**%Difference ≤2.0%**
Pass/fail	**Pass**	**Fail**	**Fail**

[a] %Difference $= 100 \times$ (injection a − injection b)/average (injection a, injection b).

that there is something wrong with the overall injection precision, even if the system suitability criterion is satisfied.

The drawback of the third approach is that the calculation is slightly time-consuming.

7.6 Sample Preparation

Although chromatographic separation is at the core of analytical method development, the sample preparation is arguably the most critical part of an analytical method. However, compared to the efforts devoted to seeking appropriate chromatographic conditions, it seems that less attention is paid to sample preparation. The sample matrix effect appears to be more of a concern in bioanalytical samples, and not in small molecule pharmaceutical products. The truth is, however, most out-of-specification results are not caused by chromatography but by incorrect sample preparation practices.

A common practice for HPLC quantitation of pharmaceutical products is to measure the UV response of an analyte in sample solutions against the UV response of an external reference standard solution. In routine analysis, the standard solution is usually prepared by dissolving the standard material in solutions without adding placebo or major ingredients of the product. The assumption is that the sample matrix does not produce any differences between the UV response of the analyte in sample solutions and the UV response of the analyte in the external standard solution that does not contain other product ingredients. For this assumption to hold, the sample matrix effect must be negligible. As we know, that, different from prescription medicinal products, the compositions of over-the-counter (OTC) drug products tend to be more complicated, due to the presence of a variety of ingredients (multiple drug substances, functional excipients, flavors, dyes, etc.). Many excipients in OTC products are meant to deliver fast onset effects, provide pleasant experiences, and improve consumer compliance. However, those excipients not only can sometimes cause product instability but also give immense challenges to analytical scientists. Some excipients interfere with the UV absorbance of the target analytes; some may not have apparent UV absorbance but can still cause inaccurate analytical results.

Li et al. have demonstrated that when an API is spiked at the same concentration into a concentrated (10×) placebo, a normal (1×) placebo, and a diluted (0.1×) placebo, respectively, the analytical results can vary dramatically [1]. Differences between placebo concentrations, autosampler temperatures, injector sample withdrawing speeds, and injection volumes on Agilent HPLC, Waters Alliance HPLC, and Waters Acquity UPLC are investigated in that study. In extreme cases, a sample response can be 90% different from the standard response, which obviously causes attention. But in many cases, the difference in responses is less than 4–5% and may be overlooked. Figure 7.8 contains the assay analyses of an API that is spiked into a 10× placebo and a 1× placebo solution. The diluent is a mixture of water and methanol. Conventional wisdom would advise us to pay attention to the sample viscosity and to lower the syringe withdrawal speed to cope with viscous sample solutions. The data points in Figure 7.8 demonstrate a distinct separation of two groups of recoveries: one is around 105%, and the other one is about 100%. The syringe withdrawal speeds do not make meaningful differences in the results. The danger of such sample matrix-induced inaccuracy in analytical results is that the product development endeavors can wrongly focus on improving product stability, content uniformity, etc. without knowing the undesired recoveries are due to unoptimized sample preparation. Furthermore, the study also shows that even with a perfect peak shape, the matrix effect can still cause false analytical results [1].

The same 1× placebo causes inaccurate analytical results, however, when the analysis is conducted on a UPLC. It seems from the study that compared to HPLC, UPLC suffers more from the matrix effect. Figure 7.9 shows that changing injection volumes does not help getting consistent recoveries if the placebo concentration is not further diluted to 0.1× level. This time the sample diluent is a mixture of water and acetonitrile, which was selected to lower the sample viscosity. Again,

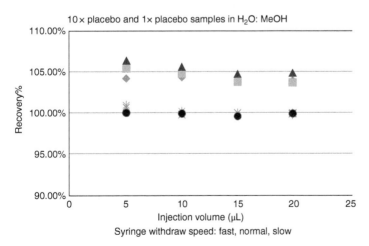

Figure 7.8 Recoveries of spiked API on an HPLC. The data center around 105% are obtained from samples that have 10x placebo and the data center around 100% are obtained from samples that have 1x placebo.

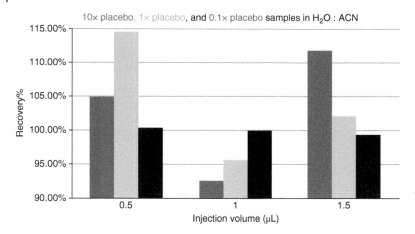

Figure 7.9 Recoveries of spiked API on a UPLC. For each injection volume, the left, middle, and right bar represents the %recovery from 10x placebo, 1x placebo, and 0.1x placebo, respectively.

the data demonstrate that it is the matrix concentration that plays a critical role in analytical recoveries.

The conclusion from this work is that the matrix effect must be taken into consideration during method development and it may not be simply ascribed to the solution viscosity, and may not be easily fixed by altering the autosampler temperature, or by using less viscous solvent, or by lowering the injector sample withdrawing speed, or by injecting less amount of sample. During method development, evaluating different injection volumes and sample withdrawing speeds to check whether consistent results could be obtained may be helpful to rule out the matrix effect. The good news is, there is a simple approach to reduce the sample matrix effect, that is, making more diluted samples. The bad news is that it may mean an increased workload as it may require second dilution during the sample preparation. Of course, detection sensitivity also needs to be considered when planning dilution.

Another sample preparation challenge the analytical scientists may face is the aging of samples. For example, the assay results of the API can be well within the expected range when testing the initial stability samples, i.e. when the product is still relatively fresh. The assay values become lower when testing three-month or six-month stability samples that are stored at higher temperatures and higher humidity conditions. The total amount of degradation products does not add up to the loss in the assay value. What needs to be done here is that during the method development, the analytical scientist should evaluate the sample extraction techniques. Samples stored at 50°C/75%RH or 60°C/75%RH for a few days to a few weeks can be used to have an early read on the sample preparation effectiveness before using the method for long-term stability studies.

7.7 Method Validation and Transfer: Mathematical Exercises or Analytical Sciences

Although each step is essential in the analytical method life cycle management, method validation is a requirement from all health authorities. There are many (regulatory) guidelines regarding the importance and approaches of method validation [2–5]. The US FDA establishes guidance for industry regarding the validation of analytical and bioanalytical methods. In 2015, the FDA revised its guideline "Analytical Procedures and Methods Validation for Drugs and Biologics," in which for the first time a requirement for method development is included. The FDA also has reviewer guidance for its officers [6]. Similarly, besides continuing to update the Q2 document that is regarding method validation, as of 2020, the ICH is also preparing a new guidance document (Q14) to provide guidance for analytical method development. International Union of Pure and Applied Chemistry published "Harmonized Guidelines for Single-Laboratory Validation of Methods of Analysis." EURACHEM, a network of organizations in Europe having the objective of establishing a system for the international traceability of chemical measurements and the promotion of good quality practices, publishes series of documents regarding the quality in analytical work [7]. The USP has several general chapters, such as <1225> "Validation of Compendial Methods" and <1226> "Verification of Compendial Methods."

Method validation is to ensure that the method is fit for its intended purpose, reproducible (rugged), and robust. It is a consensus that method validation should not be a one-time exercise to fulfill regulatory filing requirements. Regarding how to perform a method validation, such as what are the critical method parameters, how to evaluate their criticality in the method performance, etc. there are extensive discussions among academic researchers and industry [8–18]. Many pharmaceutical companies are still taking traditional approaches to conduct method validations. Table 7.7 lists the method parameters that are commonly subject to validation. Similar tables can be found in guidance documents, academic literature, or company SOPs regarding method validations.

After or included as part of a method validation, a method transfer may be carried out between a sending and a receiving laboratory. Method transfers can occur

Table 7.7 Method parameters subject to method validation.

Method parameters	Identification test	Method application		
		Potency of API (assay, content/blend uniformity, dissolution)	Potency of related compounds (impurity and degradation product)	Absence of API or limit test for related compounds
Specificity/selectivity	✓	✓	✓	✓
Sensitivity (limit of quantitation and limit of detection)	—	—	✓	✓
Working range (linearity, accuracy, recovery)	—	✓	✓	—
Precision (repeatability, intermediate precision, reproducibility)	—	✓	✓	—
Robustness (can be carried out during method development)	✓	✓	✓	✓

between an R&D laboratory and a QC laboratory at a manufacturing site; or between QC laboratories at one manufacturing site and another manufacturing site or a third party contract laboratory; or between company and company when a merger and acquisition happens. Some common practices of method transfer include comparative testing, co-validation between two laboratories or sites, full or partial method validation or revalidation, and omission of formal transfer based on the familiarity of the method in the receiving laboratory. Interestingly, however, although there are guidance documents such as USP general chapter <1224> "Transfer of Analytical Procedures," and FDA guidance for industry for method validation, detailed guidelines regarding the test parameters and acceptance criteria in method transfers are not defined in official guidance documents. Table 7.8 lists some parameters and acceptance criteria in the WHO guidelines regarding the transfer of technology in pharmaceutical manufacturing.

Table 7.8 Parameters and acceptance criteria in method transfer (WHO).

Method application	Suggested designs and acceptance criteria		
	Number of replicates	Direct comparison	Statistically derived
Assay for potency	Six independent sample preparations from one lot of homogenous product	±2% for the means; predefined range for %RSDs	Two one-sided *t*-tests with inter-site differences ≤2%, 95% confidence
Dissolution	Six or 12 units from one lot of homogenous product	±5% for the means; predefined range for %RSDs	Compare profiles (e.g. F2), or compare data at *Q* time points as for assay
Cleaning verification (recovery of residues from surfaces)	Same swabbing material is used at both sites	All samples spiked above specification should fail. At least 90% of samples spi ked below specification should pass	
Impurity, degradation, residual solvent test	Six independent sample preparations from one lot of homogenous product. Use spiked samples if necessary	The inter-site means are within ±0.05% absolute or 25% relative	Two one-sided *t*-tests with inter-site differences ≤10%, 95% confidence
Microbiological testing (qualitative and quantitative limit tests)	Triplicate	Follow acceptance limits specified in protocol	

As can be seen in Table 7.8, there are some statistical approaches in analytical method transfers. Recently, academic researchers, regulatory authorities, and nonprofit institutes such as the USP, are heavily promoting statistical approaches for method validations and transfers. Hubert et al. have published a series of scientific papers to discuss the use of statistics in method validation and transfer [15–18], which resulted in a newly established USP general chapter <1210> "Statistical Tools for Procedure Validation" in 2018.

The industry is also embracing the statistical approaches. The concept Analytical Target Profile (ATP) has gradually gained its popularity. Before starting a method development, and/or during a method validation, the goal or purpose of the method can be expressed in the format of an ATP statement. In the 2013 USP stimuli article on the life cycle management of analytical procedures, the ATP for Assay states: "The procedure must be able to quantify [analyte] in [presence of X, Y, Z] over a range of A% to B% of the nominal concentration with an accuracy and uncertainty so that the reportable result falls within ± C% of the true value with at least a 90% probability determined with 95% confidence." The ATP for impurity analysis states: "The procedure must be able to quantify [impurity] relative to [drug] in the presence of components that are likely to be present in the sample within the range from the reporting threshold to the specification limit. The accuracy and precision of the procedure must be such that the reportable result falls within ± D% of the true value for impurity levels from 0.05% to 0.15% with 80% probability with 95% confidence, and within ± E% of the true value for impurity levels > 0.15% with 90% probability determined with 95% confidence." Lots of literature work describes the approaches for method validation and transfer based on similar ATP statements.

Indeed, many publications enthusiastically discuss method validation and transfer from a purely statistical point of view, and turn the analytical activities into mathematical exercises. The percentage of organic solvents in mobile phases, flow rates, column types, mobile phase pHs, injection volumes, and column temperatures are method variables. Detection sensitivity, measurement range, method specificity, and method robustness are quality attributes of the method. The design of experiments promotes approaches to changing multiple method variables at a time to evaluate the impacts of those method variables on the method quality attributes by using statistical tools.

In reality, the majority of analytical scientists are not statisticians. Many semi-empirical approaches are still widely accepted in the industry. For example, the above statistics-heavy ATP statement can be expressed in a somewhat plainer

English that reads: "The method should be able to measure the degradation products at concentration levels between 0.05% to 0.15% with an accuracy/recovery of 90% and a precision with a %RSD less than 25% at each tested level, and should be able to measure the degradation products at concentration levels at and above 0.15% with an accuracy/recovery of 95% and a precision with a %RSD less than 10% at each tested level." To help the industry to have consistent practices, the Pharmaceutical Research and Manufacturers of America hosted workshops to discuss acceptable statistical techniques when evaluating method equivalency.

On the other hand, however, regardless of whether stating the ATP in a statistics-heavy style, or plainer English, those statements do not contain any elements of separation science. As an analytical scientist, the more important thing is to have an in-depth understanding of the analytical sciences, physical chemistry, and organic/degradation chemistry. The scientists should know what the critical quality attributes of a method really mean, and how to ensure those attributes are given enough attention from fundamental analytical and physical chemistry point of view. As mentioned previously, the statistics can only be as good as the data that are used to fit the statistical model. The scientists should do their best to understand what the data is trying to tell them, and "let the data speak for itself" rather than heavily rely on (unfamiliar) statistics.

7.8 Miscellaneous Considerations

One interesting finding in chromatography is that the baselines are noisy when the mobile phase solutions are filtered, while the baselines look much better without filtering. This observation is in contrast to the textbooks' instructions. Indeed, the author has that observation in many laboratories and thus recommends the use of HPLC grade salts and solvents and skips the filtration step. In addition, although filtering sample solutions is another textbook procedure, an analytical scientist should know that each molecule may be impacted by the filter material differently. The recoveries of all the compounds of interest should be evaluated during the filter study. In reality, many method validations only request the recovery evaluation on the API and overlook the filter effect on the degradation products. Therefore, centrifugation is a viable approach that should be evaluated so that no sample filtration is required. There are also some other observations accumulated during the years of analytical work and are summarized in Table 7.9 for readers' information.

Table 7.9 Method attributes that can bring controversies between conventional practices and real laboratory observations.

Method attribute	Conventional practice	Observations
Mobile phase filtration	Filter the mobile phases under vacuum before use, especially when they contain salts	Mobile phases are contaminated by the vacuum filtration system, resulting in noisier baselines
Sample filtration	Filter the sample solutions through syringe filters with 0.45 or 0.2 μm diameters before injection	Filtration study is only conducted on the API(s) while not on any other related compounds. Centrifugation is recommended to improve the laboratory work efficiency and to minimize the loss of related compounds during sample preparations. Guard columns maybe needed to protect the chromatographic systems
Detector setting	Follow the default setting	Attentions must be given to the following parameters: slit width, step interval, bandwidth, resolution; must be aware of the difference between: photo diode array detectors (PDA) and variable wavelength detectors (VWD), UV absorbance obtained at a pre-selected wavelength during a PDA scan and UV absorbance later extracted at a wavelength from the PDA scan
Reference wavelength	Select a wavelength such as 350 ± 100 nm	For compounds that have UV absorbance at wavelengths >290 nm, attention should be paid to the selection of reference wavelength
Binary pump vs. quaternary pump	Use interchangeably	Method robustness must be evaluated on both systems. Attentions must be paid to the solvent compressibility when using binary pump systems
Phosphate buffers at pH around 6	Use until completion	Cause peak distortion and eventually result in peak splitting if the buffer solution is used more than three days. Premix with 10% or more organic solvent can increase the mobile phase shelf life
Mobile phase pH	Adjust the pH while the pH probe is directly inserted in the mobile phase solution	Contaminations from the pH probe, especially when it was used to measure viscous samples and has not been thoroughly cleaned; the electrode storage solution can thus also be contaminated and then the probe can bring ghost peaks to the chromatograms

References

1 Li, J., Lee, J., Varanasi, M., and Xiao, K.P. (2012). Improved analytical recovery by taking into account sample matrix and chromatographic instrumentation. *American Pharmaceutical Review* 15 (7): 32–36.

2 FDA Guidance for Industry, Analytical Procedures and Methods Validation for Drugs and Biologics, July 2015.

3 ICH Q2 Validation of Analytical Procedures: Text and Methodology.

4 EMA Guideline on Bioanalytical Method Validation, February 2012.

5 WHO Guidelines on Validation – Appendix 4 Analytical Method Validation, June 2016.

6 FDA Reviewer Guidance Validation of Chromatographic Methods, November 1994.

7 (2014). *The Fitness for Purpose of Analytical Methods: A Laboratory Guide to Method Validation and Related Topics*, 2e. Eurachem.

8 Ellison, S.L.R. and Williams, A. (1998). Measurement uncertainty and its implications for collaborative study method validation and method performance parameters. *Accreditation and Quality Assurance* 3 (1): 6–10.

9 Clarke, G.S. (1994). The validation of analytical methods for drug substances and drug products in UK pharmaceutical laboratories. *Journal of Pharmaceutical & Biomedical Analysis* 12 (5): 643–652.

10 Shabir, G.A., Lough, W.J., Arain, S.A., and Bradshaw, T.K. (2007). Evaluation and application of best practice in analytical method validation. *Journal of Liquid Chromatography & Related Technologies* 30 (3): 311–333.

11 Jenke, D.R. (1996). Chromatographic method validation: a review of current practices and procedures. I. General concepts and guidelines. *Journal of Liquid Chromatography & Related Technologies* 19 (5): 719–736.

12 Jenke, D.R. (1998). Chromatographic method validation: a review of current practices and procedures. Part II. Guidelines for primary validation parameters. *Journal of Instrumentation Science & Technology* 26 (1): 1–18.

13 Jenke, D.R. (1996). Chromatographic method validation: a review of current practices and procedures. III. Ruggedness, re-validation and system suitability. *Journal of Liquid Chromatography & Related Technologies* 19 (12): 1873–1891.

14 Shabir, G.A. (2004). A practical approach to validation of HPLC methods under current good manufacturing practices. *Journal of Validation Technology*: 29–37.

15 Dewé, W., Govaerts, B., Boulanger, B. et al. (2007). Using total error as decision criterion in analytical method transfer. *Chemometrics and Intelligent Laboratory Systems* 85 (2): 262–268.

16 Rozet, E., Ceccato, A., Hubert, C. et al. (2007). Analysis of recent pharmaceutical regulatory documents on analytical method validation. *Journal of Chromatography A* 1158 (1–2): 111–125.

17 Bouabidi, A., Rozet, E., Fillet, M. et al. (2010). Critical analysis of several analytical method validation strategies in the framework of the fit for purpose concept. *Journal of Chromatography A* 1217 (19): 3180–3192.

18 Rozet, E., Ziemons, E., Marini, R.D. et al. (2012). Quality by design compliant analytical method validation. *Analytical Chemistry* 84 (1): 106–112.

Index

Analytical Scientists in Pharmaceutical Product Development: Task Management and Practical Knowledge, First Edition. Kangping Xiao.
© 2021 John Wiley & Sons, Inc. Published 2021 by John Wiley & Sons, Inc.